High-Speed (HC/HCT) CMOS Guide

High-Speed (HC/HCT) CMOS Guide

Harry L. Helms

Prentice Hall, Englewood Cliffs, New Jersey 07632

Library of Congress Cataloging-in-Publication Data

Helms, Harry L.
High-speed (HC/HCT) CMOS guide.
Includes index.
1. Metal oxide semiconductors, Complementary.
I. Title.
TK7871.99.M44H45 1989 621.3815'2 89-3754
ISBN 0-13-387929-1

Cover design: 20/20 Services, Inc.
Manufacturing buyer: Mary Ann Gloriande

The publisher offers discounts on this book when ordered in bulk quantities. For more information, write:
Special Sales/College Marketing
Prentice Hall
College Technical and Reference Division
Englewood Cliffs, New Jersey 07632

Printed in the United States of America

10 9 8 7 6 5 4 3 2 1

ISBN 0-13-387929-1

PRENTICE-HALL INTERNATIONAL (UK) LIMITED, *London*
PRENTICE-HALL OF AUSTRALIA PTY. LIMITED, *Sydney*
PRENTICE-HALL CANADA INC., *Toronto*
PRENTICE-HALL HISPANOAMERICANA, S.A., *Mexico*
PRENTICE-HALL OF INDIA PRIVATE LIMITED, *New Delhi*
PRENTICE-HALL OF JAPAN, INC., *Tokyo*
SIMON & SCHUSTER ASIA PTE. LTD., *Singapore*
EDITORA PRENTICE-HALL DO BRASIL, LTDA., *Rio de Janeiro*

This book is dedicated to the memory of A. C. Gilbert and the A. C. Gilbert Co., whose science toys and kits provided me (and many others) with their first introduction to electronics and other technical areas.

Contents

INTRODUCTION xv

1 THE ESSENTIALS OF HIGH-SPEED CMOS 1

Basics of TTL and CMOS 1
Types of HCMOS 5
HCMOS vs. Standard CMOS 6
Operating Requirements for HCMOS 8
Power Supplies 8
Interfacing Circuits 9
Adding LED Outputs 11
Open Drain Devices 12
Testing Devices for HCMOS 12
HCMOS Device Packaging 12

2 A DIRECTORY OF HCMOS DEVICES 15

74HC00 Quad 2-Input NAND Gate 16
74HC02 Quad 2-Input NOR Gate 17
74HC03 Quad 2-Input NAND Gate with Open Drain Output 18
74HC04 Hex Inverter 19

74HCT04 Hex Inverter with TTL-Compatible Inputs 19
74HCU04 Hex Unbuffered Inverter 19
74HC05 Hex Inverter with Open Drain Outputs 20
74HC08 Quad 2-Input AND Gate 21
74HC10 Triple 3-Input NAND Gate 22
74HC11 Triple 3-Input AND Gate 23
74HC14 Hex Schmitt Trigger Inverter 24
74HC20 Dual 4-Input NAND Gate 25
74HC21 Dual 4-Input AND Gate 26
74HC27 Triple 3-Input NOR Gate 27
74HC30 8-Input NAND Gate 28
74HC32 Quad 2-Input OR Gate 29
74HC34 Hex Noninverting Buffer 30
74HC42 1 of 10 Decoder 31
74HC51 4-Input and 6-Input AND–NOR Gate 32
74HC58 4-Input and 6-Input AND–OR Gate 33
74HC73 Dual J-K Flip-Flop with Reset 34
74HC74 Dual D Flip-Flop with Set and Reset 35
74HC75 Dual 2-Input Transparent Latch 36
74HC76 Dual J-K Flip-Flop with Set and Reset 37
74HC85 4-Bit Magnitude Comparator 38
74HC86 Quad 2-Input Exclusive OR Gate 39
74HC93 4-Bit Binary Ripple Counter 40
74HC107 Dual J-K Flip-Flop with Reset 41
74HC109 Dual J-$\bar{K}$ Flip-Flop with Set and Reset 42
74HC112 Dual J-K Flip-Flop with Set and Reset 43
74HC113 Dual J-K Flip-Flop with Set 44
74HC123 Dual Retriggerable Monostable Multivibrators with Resets 45
74HC125 Quad 3-State Noninverting Buffers with Inverting Enable 46
74HC126 Quad 3-State Noninverting Buffers with Noninverting Enable 47
74HC132 Quad 2-Input NAND Gate with Schmitt Trigger Inputs 48
74HC133 13-Input NAND Gate 49
74HC137 1 of 8 Decoder/Demultiplexer with Address Latch 50
74HC138 1 of 8 Decoder/Demultiplexer 51
74HC139 Dual 1 of 4 Decoder/Demultiplexer 52
74HC147 Decimal to BCD Encoder 53

74HC151 8-Input Data Selector/Multiplexer 54
74HC153 Dual 4-Input Data Selector/Multiplexer 55
74HC154 1 of 16 Decoder/Multiplexer 56
74HC157 Quad 2-Input Data Selectors/Multiplexers with Noninverting Outputs 58
74HC158 Quad 2-Input Data Selectors/Multiplexers with Inverting Outputs 59
74HC160 BCD Presettable Counter with Asynchronous Reset 44
74HC161 Binary Presettable Counter with Asynchronous Reset 61
74HC162 BCD Presettable Counter with Synchronous Reset 62
74HC163 Binary Presettable Counter with Synchronous Reset 63
74HC164 8-Bit Serial Input/Parallel Output Shift Register 64
74HC165 8-Bit Serial or Parallel Input/Serial Output Shift Register 65
74HC166 8-Bit Parallel Input/Serial Output Shift Register 66
74HC173 Quad 3-State D Flip-Flop with Common Clock and Reset 67
74HC174 Hex D Flip-Flop with Common Clock and Reset 68
74HC175 Quad D Flip-Flop with Common Clock and Reset 69
74HC190 BCD Presettable Synchronous 4-Bit Up/Down Counter 70
74HC191 Binary Presettable 4-Bit Up/Down Counter 71
74HC194 4-Bit Bidirectional Universal Shift Register 72
74HC195 4-Bit Universal Shift Register 73
74HC237 1 of 8 Decoder/Demultiplexer with Address Latch 74
74HC240 Octal 3-State Inverting Buffer 75
74HC241 Octal 3-State Noninverting Buffer 76
74HC242 Quad 3-State Bus Transceiver with Inverting Outputs 77
74HC243 Quad 3-State Bus Transceiver with Noninverting Outputs 78

74HC244 Octal 3-State Noninverting Buffer 79
74HC245 Octal 3-State Noninverting Bus Transceiver 80
74HC251 8-Input Data Selector/Multiplexer with 3-State Outputs 81
74HC253 Dual 4-Input Data Selector/Multiplexer with 3-State Outputs 82
74HC257 Quad 2-Input Data Selector/Multiplexer with 3-State Outputs 83
74HC259 8-Bit Addressable Latch/1 of 8 Decoder 84
74HC273 Octal D Flip-Flop with Common Clock and Reset 85
74HC280 9-Bit Odd/Even Parity Generator/ Checker 86
74HC283 4-Bit Binary Full Adder with Fast Carry 87
74HC299 8-Bit Bidirectional Universal Shift Register with Parallel Input/Output 88
74HC354 8-Input Data Selector/Multiplexer with Data and Address Latches and 3-State Outputs 89
74-HC356 Clocked 8-Input Data Selector/ Multiplexer with Data and Address Latches and 3-State Outputs 91
74HC365 Hex 3-State Noninverting Buffer with Common Enables 92
74HC366 Hex 3-State Inverting Buffer with Common Enables 93
74HC367 Hex 3-State Noninverting Buffer with Separate 2-Bit and 4-Bit Sections 94
74HC368 Hex 3-State Inverting Buffer with Separate 2-Bit and 4-Bit Sections 95
74HC373 Octal 3-State Noninverting Transparent Latch 96
74HC374 Octal 3-State Noninverting D Flip-Flop 97
74HC377 Octal Noninverting D Flip-Flop with Output Enable 98
74HC386 Quad 2-Input Exclusive OR Gate 99
74HC390 Dual 4-Stage Binary Ripple Counter with Divide-by-Two and Divide-by-Five Sections 100
74HC393 Dual 4-Stage Binary Ripple Counter 101

74HC533 Octal 3-State Inverting Transparent Latch 102
74HC534 Octal 3-State Inverting D Flip-Flop 103
74HC540 Octal 3-State Inverting Buffer/Line Driver/Line Receiver 104
74HC541 Octal 3-State Noninverting Buffer/Line Driver/Line Receiver 105
74HC563 Octal 3-State Inverting Transparent Latch 106
74HC564 Octal 3-State Inverting D Flip-Flop 107
74HC573 Octal 3-State Noninverting Transparent Latch 108
74HC574 Octal 3-State Noninverting D Flip-Flop 109
74HC589 8-Bit Serial or Parallel Input/Serial Output Shift Register with 3-State Output 110
74HC595 8-Bit Serial Input/Serial or Parallel Output Shift Register with Latched 3-State Outputs 112
74HC597 8-Bit Serial or Parallel Input/Serial Output Shift Register with Input Latch 114
74HC640 Octal 3-State Inverting Bus Transceiver 116
74HC643 Octal 3-State Inverting and Noninverting Bus Transceiver 117
74HC646 Octal 3-State Bus Transceiver and D Flip-Flops with Noninverting Outputs 118
74HC648 Octal 3-State Bus Transceiver and D Flip-Flops with Inverting Outputs 120
74HC651 Octal 3-State Bus Transceiver and D Flip-Flops with Inverting Outputs 121
74HC652 Octal 3-State Bus Transceiver and D Flip-Flops with Noninverting Outputs 123
74HC688 8-Bit Equality Comparator 124
74HC4002 Dual 4-Input NOR Gate 125
74HC4016 Quad Analog Switch/Multiplexer/Demultiplexer 126
74HC4017 Decade Counter 127
74HC4020 14-Stage Binary Ripple Counter 128
74HC4024 7-Stage Binary Ripple Counter 129

74HC4040 12-Stage Binary Ripple Counter 130
74HC4046A Phase-Locked Loop with Voltage-Controlled Oscillator 131
74HC4049 Hex Inverting Buffers/Logic-Level Down Converters 132
74HC4050 Hex Noninverting Buffers/Logic-Level Down Converters 133
74HC4051 Single-Pole, 8-Position Analog Multiplexers/Demultiplexers with Common Off 134
74HC4052 Double-Pole, 4-Position Analog Multiplexers/Demultiplexers with Common Off 136
74HC4053 Triple Single-Pole, Double-Position Analog Multiplexers/Demultiplexers with Common Off 137
74HC4060 14-Stage Binary Ripple Counter with Oscillator 138
74HC4066 Quad Analog Switch/Multiplexer/Demultiplexer 139
74HC4075 Triple 3-Input OR Gate 140
74HC4078 8-Input NOR/OR Gate 141
74HC4511 BCD to Seven-Segment Latch/Decoder/Display Driver 142
74HC4514 1 of 16 Decoder/Demultiplexer with Address Latch 144
74HC4538 Dual-Precision Retriggerable, Resettable Monostable Multivibrator 146
74HC4543 BCD to Seven-Segment Latch/Decoder/Display Driver for LCDs 122

3 BASIC LOGIC BUILDING BLOCKS

Combinational and Sequential Logic 151
Fundamental Types of Logic Gates 152
The 74HC00 Quad NAND Gate 154
Flip-Flop Circuits from Logic Gates 155
Using Other Logic Gate Devices 159
Applications of Logic Gates 161

4 APPLICATIONS OF THE 74HCU04 HEX UNBUFFERED INVERTER 167

Oscillator Applications 167
Amplifier Applications 170
Interfacing and Drive Applications 171

5 CLOCKED LOGIC ELEMENTS 173

Clocking Methods 173
Generating Clock Signals 175
Flip-Flop Devices 175
Counter/Dividers and Their Applications 179
Shift Registers 184
Other Clocked Logic Applications 185

6 TRI-STATE OUTPUT DEVICES 189

Basic Tri-State Devices 190
Bus Interfacing Devices 192
Bus Transceivers 193
Bus Storage Registers 194
Other Tri-State Logic Devices 198

7 MISCELLANEOUS HCMOS DEVICES 199

The 74HC4538 Dual Monostable Multivibrator 199
The 74HC280 Parity Generator/Checker 200
Equality and Magnitude Comparison 202
Analog Multiplexers/Demultiplexers 204
Analog Switches 207

INDEX 209

Introduction

The effort to find the best all-around logic family has been a long and costly one.

The first entrants were the RTL and DTL families. While it was nice to have various digital logic functions available on a single chip, such IC families were idiosyncratic and cranky, especially when used in complex circuits. Something better was needed.

RTL and DTL logic were blown away with the introduction of the TTL (*transistor-transistor logic*) family. Numerous TTL devices were introduced, providing a variety of logic functions. TTL was fast, and its outputs could serve as the inputs for several other TTL devices. And it was cheap.

But TTL was not perfect. TTL required a +5 V power supply, and a TTL device could easily be "fried" if this were exceeded by even 0.5 V. TTL was also a "power hog." TTL circuits consumed large amounts of current, typically 3 mA per gate, making it a poor choice for battery-powered applications. The large current consumed by TTL also made it a "noisy" logic family. When TTL gates switched states, the resulting current demands placed on the supply produced "spikes" and variations in the supply voltage, which could cause false triggering of the devices and erratic operation of the entire circuit. Decoupling capacitors and similar measures had to be taken to minimize such effects. And TTL was fussy about clean logic-level transitions in clock signals and other inputs.

The battle really got hot with the debut of CMOS (*complementary metal-oxide semiconductor*) logic devices. Unlike TTL, CMOS could be operated from supply voltages ranging from +3 to 15 V. Moreover, the current demands of CMOS were much less than those of TTL—less than 0.1 mA per gate—making battery-powered logic circuits practical. CMOS was also a much more "coöperative" logic family than TTL. It was less bothered by noise, could operate with less-than-perfect input and clock signals, and did not need decoupling capacitors and similar precautions.

But CMOS also had problems that TTL did not. One that vexed many users was its susceptibility to damage from static electricity, which forced one to take several annoying precautions in the storage and installation of CMOS ICs. More important for many, however, was CMOS's serious lack of speed compared with TTL. TTL was a hare and by comparison CMOS was a tortoise. And this was one race the hare won going away.

TTL did not take the CMOS challenge lightly. It struck back with LS (*low-power Schottky*) and ALS (*advanced low-power Schottky*) versions of TTL. These offered significantly reduced power consumption than standard TTL, although still more than CMOS, while maintaining TTL's high operating speed. Unfortunately, they still suffered from other TTL shortcomings and cost more than standard TTL devices.

So those were the choices, and deciding on which logic device to use involved several trade-offs. If you wanted speed alone, there was TTL or one of its variations. If you wanted a wide supply voltage range or low current consumption, CMOS was the answer. But if you needed a logic family that combined the best of TTL and CMOS with as few of their shortcomings as possible, then you had a problem. Often the only real choice was to combine TTL and CMOS devices in the same circuit and put up with two different supply voltages (unless one operated CMOS at +5 V, which severely degraded its performance) and the necessity to interface the two device families.

Until recently, that is.

The search for the best of all possible "logic worlds" may be over. And *high-speed (HC/HCT) CMOS* is the answer. Just imagine that the king and queen of the two warring factions, TTL and CMOS, married and produced offspring with as many of their strong points as possible and few of their weaknesses. That's high-speed CMOS (or HCMOS as we will refer to it)—high speed *and* lower power consumption.

HCMOS is considerate of those who have spent time learning how to use other logic families, particularly TTL. Virtually every TTL device has an HCMOS equivalent, with identical pin connections and functions. For example, the TTL 7400 quad 2-input NAND gate IC has an HCMOS equivalent known as the 74HC00. The HCT versions of HCMOS (such as the 74HCT00) are "drop-in" replacements for existing TTL devices and can freely be mixed in the same circuit as TTL so long as a +5 V supply voltage is used. Moreover, HCMOS equivalents of popular CMOS devices whose functions are not duplicated in TTL, such as the 4046 phase-locked loop, are now available and more are being added each month. No other device family offers so many different devices.

HCMOS is not perfect. The operating voltage range is less than that of the standard CMOS, and decoupling capacitors are frequently necessary. ALS versions of TTL are still slightly faster. HCMOS can be damaged by static electricity. And HCMOS is, at the time this book is being written, more expensive than standard TTL and CMOS. Efforts to develop a more nearly perfect logic family will continue, but HCMOS is clearly the wave of the future. Major semiconductor companies such as Motorola, Texas Instruments, GE/RCA Solid State, and Signetics have introduced extensive lines of HCMOS ICs. Other companies, and more devices, are certain to follow.

This book aims to be both an introduction to HCMOS and a continuing reference. You will find an introduction to HCMOS and its operating conditions, descriptions and pin connections for the most popular HCMOS devices, and applications of popular HCMOS devices. And while this book is not intended as an introduction to digital logic, tutorial material has been included to help you master some principles of logic circuits in case it is your first experience with digital ICs. Since HCMOS will likely be the most important standard logic family during the 1990s, it makes sense to start learning about digital logic with it.

That's enough of an introduction. Let's get going. . . .

Harry L. Helms

High-Speed (HC/HCT) CMOS Guide

1

The Essentials of High-Speed CMOS

What are the advantages of HCMOS over standard CMOS or TTL? Here are some of the most important:

- Much lower power consumption than TTL
- Better noise immunity
- Wider operating voltage range than TTL (for HC devices)
- Outputs can directly interface to CMOS or TTL devices
- Wider operating temperature range
- Lower input current
- Higher reliability
- Greater variety of available devices, including functional equivalents for virtually all TTL and many CMOS ICs

That is an impressive list. In this chapter, we will look at some of the fundamentals of HCMOS and lay the foundation for making use of the remainder of this book.

BASICS OF TTL AND CMOS

Since HCMOS is so frequently referenced to and discussed in terms of the existing TTL and CMOS logic families, we should review some of the fundamental concepts and terminology of those families to better understand how

HCMOS devices work. We will examine them through NAND gates implemented using both technologies. A NAND gate normally has a high logic-level output unless all of its inputs are high; when this happens, the output goes low. NAND gates are the basic "building blocks" of digital electronics, since *any* digital or logic can be implemented using nothing more than NAND gates.

Figure 1-1 shows a TTL NAND gate. Like all TTL family devices, this circuit requires a DC supply voltage (known as Vcc) of +5 V, with a tolerance of 5%. TTL devices will recognize +5 V as a high (or 1) logic level, and they will generally recognize any voltage in excess of +2.4 V as a high logic-level signal. TTL devices will also recognize signals of +0.8 V or less as low (or 0) logic levels. However, the normal practice is to use +5 V for high levels and 0 V (known as ground) for low levels.

As you can see, the TTL NAND gate in figure 1-1 uses bipolar transistors. One of them, labeled Q1, has two emitters. Q1 functions much like two transistors with their bases and collectors connected in parallel. When either Input 1 or Input 2, or both, have a low input, Q1 is saturated. Q2 is then turned off and the resulting output is high. If both Input 1 and Input 2 are high, Q1 acts as an inverted transistor; its base-collector junction acts as a conducting diode. This turns Q2 on, and the resulting output is low. Note that there are two diodes across the inputs of this circuit. These are known as *input clamping* diodes. Their purpose is to prevent "ringing" during sharp rises in the input signals at high frequencies, which improves the device's immunity to noise in the input signal.

One significant aspect of TTL is that any of its inputs will source current into the output or device driving it. For standard TTL devices, this is typically 1 mA. Thus, to drive a TTL input low, current must be sunk.

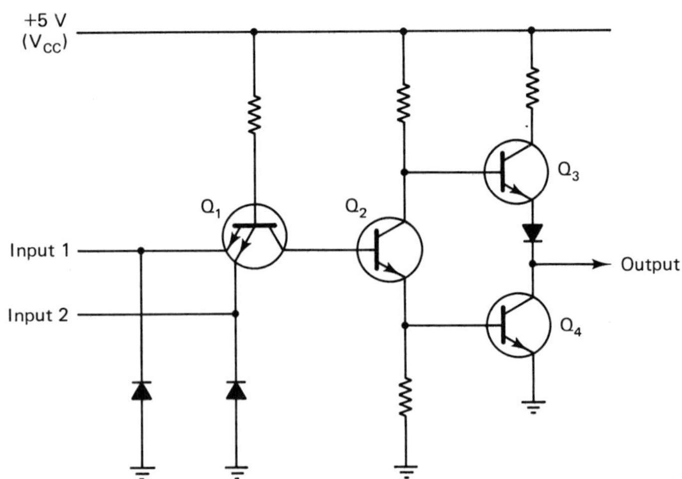

Figure 1-1 Schematic Equivalent of TTL NAND Gate

The speed of a TTL device is measured in kHz or MHz, and a single "Hertz" can be defined as switch in the output state of a single gate (from low to high or vice versa). Several TTL devices are capable of operating at such high rates as 25 to 50 MHz. The rate at which a change in a gate's inputs is reflected in the output is known as *propagation delay.* This can range from 10 nanoseconds (ns) for standard TTL to 3 ns or less for TTL varieties optimized for speed.

The number of inputs a single output can drive is known as *fanout.* A single standard TTL output can drive up to 10 standard TTL or 20 LS (low-power Schottky) TTL inputs. A single LS TTL output can drive up to 5 standard TTL or 10 LS TTL inputs.

Perhaps the greatest single shortcoming of TTL lies in its current consumption. For standard TTL devices, this is typically 8 mA *per gate.* While advanced versions of TTL have improved this figure, TTL logic is still significantly inferior to other families (such as CMOS) in this respect. Not only does this make battery-operated circuits less practical, but it also creates large current "spikes" throughout the power supply when a gate changes local levels. These can cause irregular operation if the spikes are mistaken for input signal changes. To prevent this, decoupling capacitors should be connected across Vcc to ground of each TTL device in a circuit.

Figure 1-2 shows a CMOS NAND gate. The most immediate difference you will notice between this circuit and the one in figure 1-1 is in the transistors used. The circuit in figure 1-1 uses standard bipolar transistors, while the circuit in figure 1-2 uses *metal oxide semiconductor* (MOS) transistors. These are of the field effect transistor (FET) variety, commonly known as

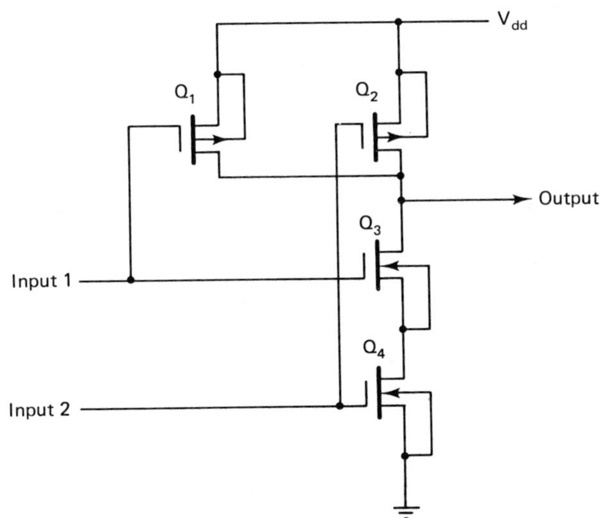

Figure 1-2 Schematic Equivalent of CMOS NAND Gate

MOSFETs. Another name for them is *insulated gate* FETs. Unlike bipolar transistors, these have four terminal connections. Three of these have rough equivalents to the three terminals of a bipolar device: the source (similar to a bipolar emitter), gate (base), and drain (collector). The fourth terminal is the *body*, or *substrate*. This is the terminal indicated by an arrow. In operation, a MOSFET has the gate area "insulated" from the conducting path (such as the channel between the drain and the source) by a thin layer of glass. The gate affects the operation of the drain-source circuit through an electric field it develops, not through a direct electrical contact. The body section of a MOSFET must be held to a nonconducting voltage level, which can be either positive or negative depending on the type of MOSFET involved. The two types of MOSFETs are known as *n-channel* and *p-channel,* which are roughly equivalent to NPN and PNP bipolar transistors, respectively. These are identified by the direction of the "arrow" in the schematic symbol; an "inward" arrow indicates an n-channel MOSFET, while an "outward" arrow designates a p-channel device.

One advantage of MOSFETs over bipolar transistors is that many of them can be constructed in a small area, making them an excellent choice for integrated devices. Figure 1-2 shows a CMOS NAND gate constructed using *complementary* MOSFETs. The term "complementary" comes from the fact that the n-channel and p-channel transistors serve, in effect, as each other's load resistors. In operation, both Input 1 and Input 2 must be held at a high logic level to turn on the complementary transistor pair Q3 and Q4. This also turns off transistors Q1 and Q2.

The supply voltage for CMOS is known as V*dd*. This supply voltage can vary over a wide range, typically +3 to +18 V. "Standard" CMOS voltages commonly used in actual circuits are +5 and +15 V. The advantages of CMOS are more pronounced at higher voltage levels. CMOS also has no input current, and its power consumption is almost zero unless the gate or device is actually switching states. The exact power consumption increases with operating speed of the device, but a typical figure is 1 μA per gate at 1 kHz. As with TTL, the convention is to use the supply voltage to represent a high logic level and 0 V for a low level, However, a voltage between 0 V and approximately half the supply voltage will be recognized as a low logic level. If high logic is Vdd and low is 0 V, this means that CMOS will not mistake transient noise signals as high logic signals. (It also means that the noise immunity of CMOS is further enhanced by using higher supply voltages.) The low-current consumption of CMOS does not, as a general rule, produce "spikes" in the power supply system and makes the use of decoupling capacitors unnecessary.

Fan out capability of CMOS is greatly improved over TTL. Most CMOS devices can drive over 50 other CMOS ICs.

CMOS has a few significant shortcomings, however. One is its slower operating speed. The maximum speed of most standard CMOS devices is less than 5 MHz, and power consumption rises markedly at higher operating rates. Propagation delay is also greater, with 125 ns being a typical figure. At the upper frequency range of most CMOS devices, their power consumption is nearly equal to an equivalent TTL device.

Perhaps the most serious fault with CMOS is its high susceptibility to damage from static electricity. This is because the construction of the integrated MOS transistors form what are, in effect, capacitors; static electricity can "puncture" the insulating glass keeping the gate area from the drain-source channel. This renders the integrated transistor inoperative. As a result, special precautions must be taken to prevent buildup and discharge of static charges when storing and handling CMOS devices and when building and using circuits containing them.

TYPES OF HCMOS

HCMOS devices are identified by the letters "HC," "HCT," or "HCU" in their device numbers, which begin with "74" or "54/74." HCMOS device numbers are based on their TTL or CMOS functional equivalents. For example, "74HC00" refers to the HCMOS equivalent of the TTL 7400 quad 2-input NAND gate, while "74HC4046A" is the HCMOS equivalent of the CMOS 4046 PLL IC. Each identifying code—"HC," "HCT," and "HCU"—indicates a specific type of HCMOS.

The "HC" designation indicates what might be termed a "standard" HCMOS device. It is capable of operating over a supply voltage range of +2 to 6 V DC. HC devices have CMOS input switching levels and buffered CMOS outputs. They can accept CMOS inputs, and their outputs can directly drive CMOS, TTL, or NMOS (n-channel MOS) devices.

The "HCT" designation indicates an HCMOS IC with a TTL to CMOS input buffer stage to enable it to directly accept TTL inputs. *Unlike HC devices, HCT ICs must have a supply voltage restricted to +5 V DC, the same as TTL devices.* In many cases, these HCT ICs may be directly substituted for TTL and LS/TTL devices in existing circuit designs. Their outputs can directly drive HCMOS or TTL devices.

The "HCU" designation is, at the time of this writing, used with only one device—the 74HCU04 hex inverter. "HCU" indicates an *unbuffered* device suitable for linear applications. Input and output levels may differ from those of buffered devices. Outputs can directly drive CMOS, TTL, or NMOS. Applications of the 74HCU04 will be discussed later in this book.

While HC and HCT versions of HCMOS are functionally equivalent, *they are not operationally equivalent.* Compared with a comparable HC device, HCT devices have greater power consumption, have lower noise immunity, and, as mentioned before, are restricted to operating at a +5 V supply voltage—in many respects, HCT versions of HCMOS have more in common with

LS versions of TTL than they do with HC HCMOS ICs. This means that, in most cases, HCT devices can be used as direct replacements for TTL ICs. Whether to use HC or HCT devices in a circuit—or whether to mix the two together—is the source of much dispute. The consensus seems to be emerging that "all-HC" circuits are preferred, since this will give the advantages of HC ICs (power consumption, noise immunity, etc.); HCT devices should be used with HC ICs *only* for logic-level (TTL to CMOS) conversions.

Note the "Vcc" is used to denote the supply voltage for HCMOS devices even though HCMOS devices are not restricted to +5 V. But +5 V has emerged as the "standard" operating voltage for HCMOS, largely because the numerous power supply circuits developed for TTL can be used for HCMOS with no modification. A secondary standard voltage is +3 V, which can be obtained from two dry cell batteries and is becoming popular for portable circuits using HCMOS.

HCMOS VS. STANDARD CMOS

The differences between standard CMOS and HCMOS begin with the fabrication process and involve the substance used to form the gate section of each integrated MOSFET. Standard CMOS uses a metal gate process, meaning that the transistors fabricated on the silicon wafer have metal gates, while HCMOS uses a *silicon gate* process. The silicon gate process uses the gate to define the channel during IC fabrication, eliminating registration errors and the need for gate overlaps. The silicon gate process allows smaller gate and channel lengths, giving HCMOS a substantial space savings in device size—a typical HCMOS device is about half the size of a conventional metal gate CMOS IC. The smaller gate length also means that HCMOS has a more efficient use of chip area and higher drive capability per gate width. The elimination of gate overlap greatly reduces gate capacitance and increases speed capability.

The differences between metal gate and silicon gate processing show up in HCMOS and standard CMOS specifications. Here are some of the more significant ones:

Operating Voltage Range

Standard CMOS: +3 to 18 V DC
HCMOS: +2 to 6 V DC

Typical Power Consumption

Standard CMOS: 1 μA per gate (1 kHz)
HCMOS: 5 μA per gate (1 kHz)

Operating Frequency (D Flip-Flop)

Standard CMOS: 4 MHz
HCMOS: 40 MHz

Propagation Delay

Standard CMOS: 125 ns
HCMOS: 8 ns

Output Interfacing to TTL

Standard CMOS: requires interfacing for TTL loads
HCMOS: can directly drive up to 10 TTL loads

There is an unfortunate similarity between standard CMOS and HCMOS in their mutual susceptibility to damage from static discharge. The gate oxide for HCMOS breaks down at a gate-source potential of about 100 V. An input protection network is incorporated into HCMOS devices; the schematic for this is shown in figure 1-3. This network generally offers improved protection from static damage compared with standard CMOS devices. However, HCMOS is still clearly inferior to TTL devices in this respect. To prevent damage, the following handling precautions for HCMOS should be observed:

- HCMOS ICs should be stored in materials that are conductive or antistatic, such as conducting foam, when stored or transported; nonconductive plastic foam or "snow" must be avoided.
- Place devices with pins down on aluminum foil sheets or trays when not in a circuit or being stored.
- Use a battery-powered soldering iron to solder HCMOS device pins.
- Connect low impedance devices (pulse generators, etc.) to HCMOS device inputs only after supply voltage is applied to the device.

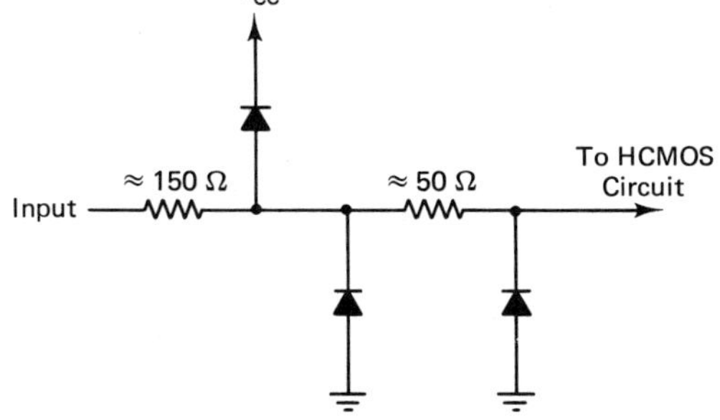

Figure 1-3 Input Protection Circuit for HCMOS Devices

- Ground yourself by touching a metal surface or grounding strap before handling HCMOS devices.
- Handle circuit boards containing HCMOS devices with the same care, since they are "extensions" of the ICs.

OPERATING REQUIREMENTS FOR HCMOS

The basic operating requirements for HCMOS are simple and very similar to those for standard CMOS. Among these are:

- Use the supply voltage (Vdd) to represent high logic signals and 0 V (ground) to represent low logic signals.
- Input voltages must not exceed Vdd.
- All unused inputs must be connected to Vdd or ground, as appropriate, to prevent erratic operation of the device.
- All unused outputs should be left floating (i.e., open).
- Never apply an input signal to a device unless supply voltage is being applied to the device.

The rapid switching of gates and increased power consumption of HCMOS means that decoupling capacitors may be required, especially at higher operating frequencies and in certain circuits. Values should be between 0.022 and 0.1 μF. Each circuit board containing HCMOS ICs should be bypassed. In addition, bypassing capacitors should be placed across all synchronous counters, devices having oscillator elements, and Schmitt trigger devices.

POWER SUPPLIES

While HCMOS devices can operate from a power source ranging from +2 to 6 V, it has become standard practice to use +5 V supplies. This is the same voltage level used for TTL devices, and as a result any of the popular power supplies developed for use with TTL circuits can be used for HC, HCT, and HCU devices as well.

The lower power consumption of HCMOS devices makes battery power supplies practical for many applications. Figure 1-4 shows a circuit that will reduce a +6 V battery supply to approximately +5 V. The only crucial part of this circuit is the polarity of the two 1 μF electrolytic capacitors.

A useful AC power supply for most "bench" applications is shown in figure 1-5. This circuit takes 110–120 V AC power line current and provides +5 V DC at slightly over one ampere of current, which should be more than

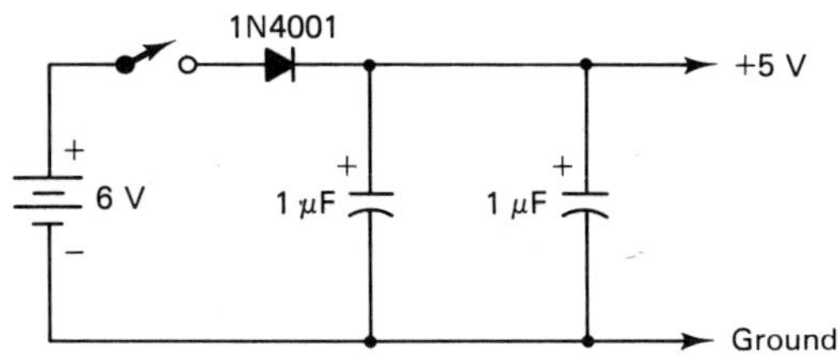

Figure 1-4 Five-Volt Battery Power Supply for HCMOS Devices

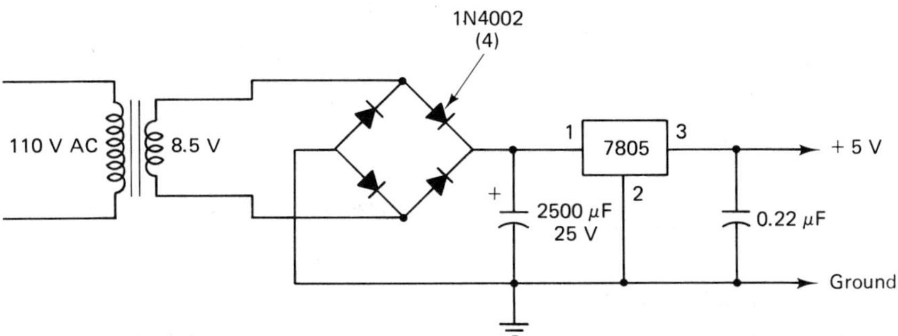

Figure 1-5 Five-Volt AC Line Power Supply for HCMOS Devices

enough for most applications. The transformer used should have a 110–120 V primary and an 8.5 V secondary; the transformer output is then rectified by the bridge circuit consisting of four 1N4002 diodes. Voltage regulation is provided by a 7805 voltage regulator IC.

INTERFACING CIRCUITS

As mentioned earlier, one of the major advantages of high-speed CMOS is that it has functional equivalents to virtually all TTL and a growing number of standard CMOS devices. There is rarely a genuine need to use a non-HCMOS device. However, it is often necessary to do some interfacing, as when a circuit board or card using HCMOS must be interfaced to another card or board that uses TTL or standard CMOS. Fortunately, HCMOS devices have sufficient output current devices to directly drive TTL or standard CMOS inputs so long as both use the same supply voltage. Moreover, HCT versions of HCMOS can be directly driven from TTL (although, as mentioned in the previous section, good practice is to use HCT only for interfacing and level conversion applications). Figure 1-6 shows these basic interfacing techniques.

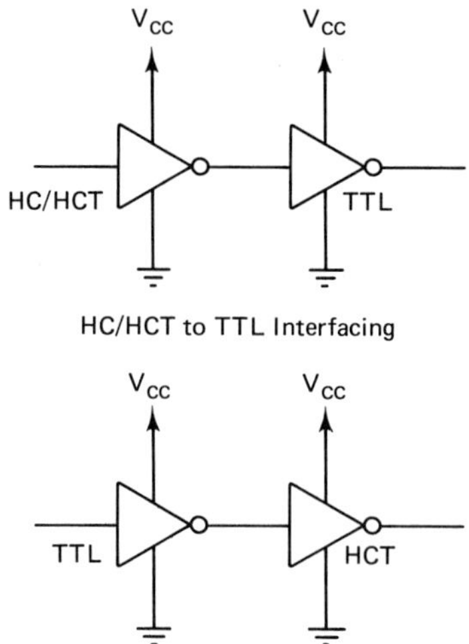

Figure 1-6 HC/HCT to TTL and TTL to HCT Interfacing Methods

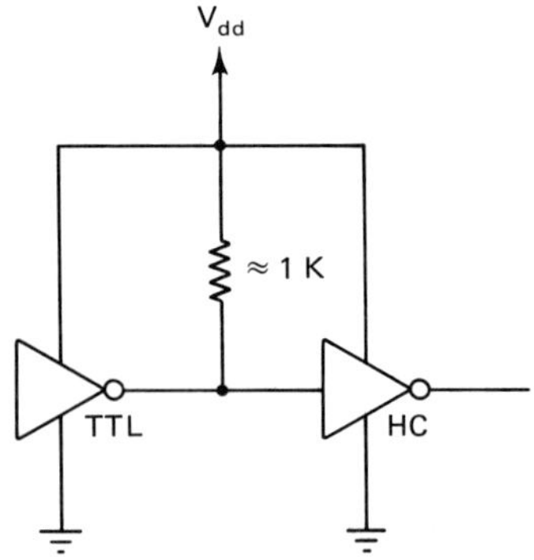

Figure 1-7 TTL to HC Interfacing Method

The output of a TTL device can drive an HC device input if a "*pull-up*" resistor is used, as shown in figure 1-7. The value of the pull-up resistor will depend on the TTL device's output current, but a value of approximately 1 K will usually suffice. This is not a preferred approach, with the TTL to HCT method shown in figure 1-6 favored.

Sometimes it may be necessary to interface HCMOS to TTL or standard CMOS devices operating at different voltage levels. Figure 1-8 shows how to interface standard CMOS operating at high-voltage levels to HCMOS operating at a lower-voltage level. A typical situation may be standard CMOS operating at +15 V interfaced to HCMOS operating at +5 V. The necessary level conversion is performed by one section of either a 75HC4049 hex inverting buffer device or by a 74HC4050 hex noninverting buffer IC. These are HCMOS devices and are powered by the same supply voltage used by the HCMOS it interfaces to the standard CMOS device.

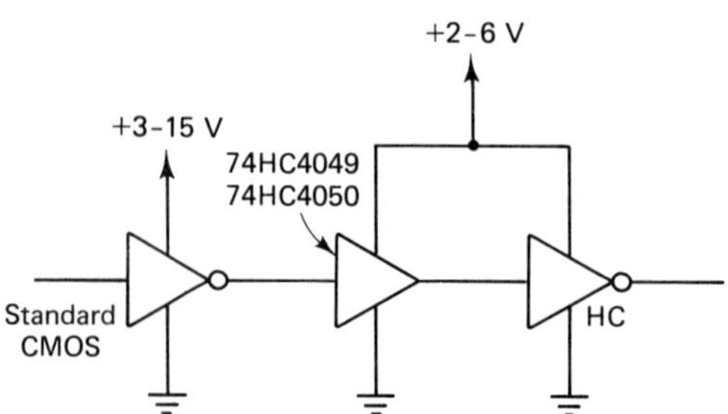

Figure 1-8 Interfacing High-Voltage Standard CMOS to Lower-Voltage HCMOS

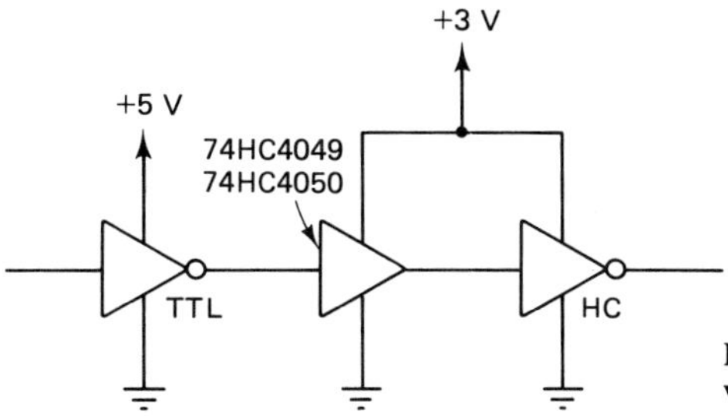

Figure 1-9 Interfacing TTL to Low-Voltage HCMOS

In some situations, HCMOS may be used at a lower voltage level than a TTL device that supplies an input signal, as shown in figure 1-9. Like figure 1-8, this also uses either a 74 HC4049 or a 74HC4050 to perform the level conversion. Both the HC IC and the hex buffer are powered by the same supply voltage (+3 V in this example), while the TTL device operates from +5 V.

It should again be emphasized that interfacing is to be avoided and kept to the absolute minimum necessary. To mix HCMOS devices in the same circuit as standard CMOS or TTL merely defeats the purpose of using HCMOS in the first place, namely, to overcome the disadvantages of other logic families. The best approach is to go with a purely HCMOS circuit design unless mixing device families and voltages is unavoidable.

ADDING LED OUTPUTS

Light-emitting diodes (LEDs) make ideal output indicators for logic circuits, since they can quickly react to changes in logic levels and are easily read: A lit LED indicates a high logic signal, while an unlit (extinguished) LED indicates a low logic signal. Figure 1-10 shows how to add LEDs to HCMOS devices. The 1 K resistor used between the HCMOS signal and LED is known as the *current limiting* resistor and is used to prevent damage to the LED.

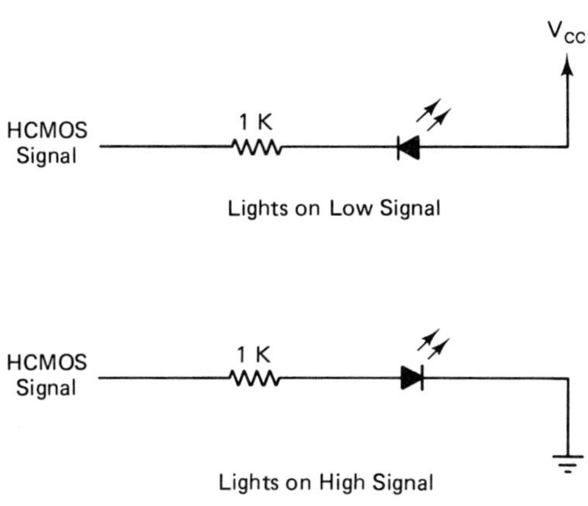

Figure 1-10 Adding LEDs to HCMOS Device Outputs

OPEN DRAIN DEVICES

Some HCMOS ICs are designated as *open drain output* devices. This means that these devices are specifically designed to sink current to ground. When used with suitable pull-up resistors, these devices can be used in so-called wired applications where several outputs are connected together. Other applications include any that requires a sinking current, such as LED drivers.

TESTING DEVICES FOR HCMOS

The most useful testing device for HCMOS is the old reliable *logic probe* familiar to anyone who has worked with TTL or standard CMOS circuits. A logic probe is a hand-held instrument with a pointed "probe" at one end which is brought into contact with the input or output of a digital logic device. Most logic probes use LEDs to indicate high or low logic levels at the point the probe touches; some also have a separate LED to indicate a pulsing (changing) logic state at the point under test. Almost all logic probes can be used with either TTL or CMOS devices, and these are ideal for use with HCMOS. Most logic probes are powered from the circuit under test; if not, they should use the same supply voltage as the HCMOS device being tested.

A useful companion to a logic probe is a *logic pulser,* or *logic injector.* This device inputs substitute logic pulses into a circuit. As with logic probes, most of these can be used with either TTL or CMOS devices and work well with HCMOS. The output pulses will typically be square waves, and output frequencies can be as high as 500 Hz. Some logic pulsers can output a single pulse if desired; it is also possible to build "single pulse injectors" from HCMOS devices if desired.

Of course, it is also possible to use oscilloscopes to display input and output levels of HCMOS devices and circuits. For most purposes, however, this is overkill; a good logic probe and logic pulser will suffice for most HCMOS testing needs.

HCMOS DEVICE PACKAGING

Most HCMOS devices are available in three major types of dual in-line (DIP) packages; *plastic, ceramic,* and *surface mount.* The first two types come with conventional IC pins that can be soldered, wire-wrapped, or inserted into sockets and solderless "breadboards." Surface mount devices, by contrast, come with short, flattened "pins" designed for mounting and soldering on the same side of the circuit board. From a functional standpoint, there is no difference between packages.

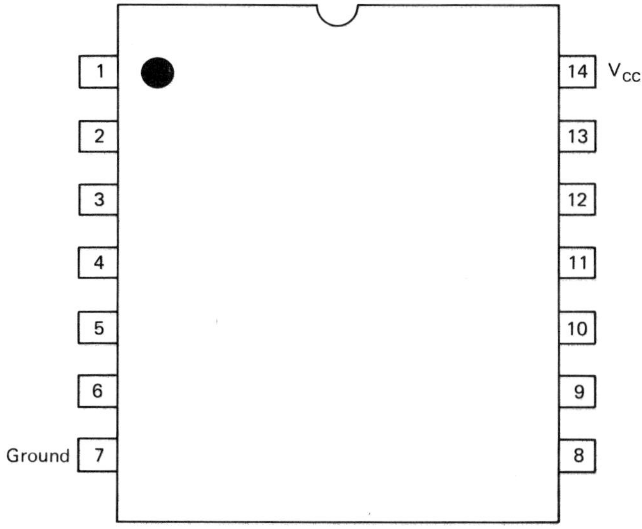

Figure 1-11 Dual In-Line Package (DIP) IC Packaging and Pin Arrangement

Regardless of the package type or the exact number of pins, pins are numbered according to the pattern shown in figure 1-11. Most packages will have an indentation or notch to indicate the "top" of the package. Most will also have a dot or other marking to denote the first pin of the device, as shown in figure 1-11. By convention, the last pin in the lower-left corner of the device is for the ground connection, and the last pin in the upper-right corner is for the supply voltage.

The type of packaging used can be determined by a letter suffix attached to the device part number. For devices supplied by Motorola and Signetics, a *J* suffix is used for ceramic packages, an *N* suffix for plastic packages, and a *D* suffix for surface mount packages. For devices supplied by GE/RCA Solid State, *F* indicates a ceramic package, *E* is used for plastic, and *H* indicates a surface mount device.

2

A Directory of HCMOS Devices

New HCMOS devices are constantly being added. At the time this book is being written, virtually all TTL devices have HCMOS equivalents. Moreover, an increasing number of HCMOS equivalents of popular standard CMOS devices are becoming available. Standard CMOS equivalents are being added selectively, and those types of standard CMOS functions that are already adequately covered by TTL equivalents are not likely to appear soon in HCMOS versions.

This chapter contains a directory of the most popular and interesting HCMOS devices currently available, although it is not all-inclusive. All the devices share the general specifications and characteristics of HCMOS discussed in the first chapter. For each device, we have given pin connections, pin functions, an internal diagram (except where impractical), a logic or function table, and a brief description of the device and its operation. More information on using these devices will be given in subsequent chapters of this book.

In the logic tables accompanying most device descriptions, "H" indicates a high logic level, "L" indicates a low logic level, and "X" indicates that the logic level is irrelevant to the operation of the device. (This is sometimes known as the "don't care" condition.) For clock signals, "F" denotes a clock signal that changes from a high to a low logic state (a "falling" or negative-edge signal), while "R" indicates a clock signal that changes from a low to a high logic level (a "rising" or positive-edge signal). If the output is a pulse rather than a steady logic signal, "HP" denotes a high-level pulse, while "LP" denotes a low-level pulse. A "Z" indicates that the output displays a high impedance which effectively "disconnects" the output from the rest of the circuit. A horizontal bar (‾) over a label or identifier indicates that it is the *complement* of the normal state; for example, if the normal state is high, its complement is low. Other abbreviations and symbols used in a particular table will be defined as needed in the device description.

1 74HC00 Quad 2-Input NAND Gate

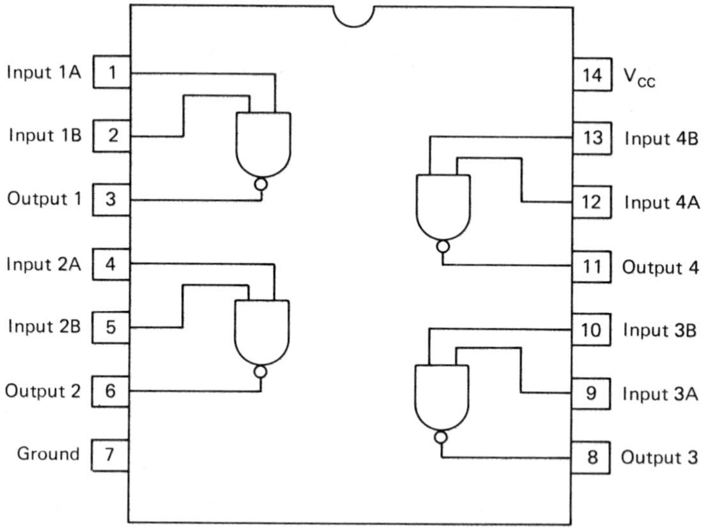

This device contains four independent NAND logic gates. Each gate consumes less than 2 μA of current while in the quiescent mode. Input current is typically less than 1 μA. The propagation delay decreases as the supply voltage increases; it is 90 ns at +2 V and 18 ns at +5 V. Input capacitance is 10 pF. The truth table for each gate is:

INPUTS		OUTPUT
L	L	H
L	H	H
H	L	H
H	H	L

2 74HC02 Quad 2-Input NOR Gate

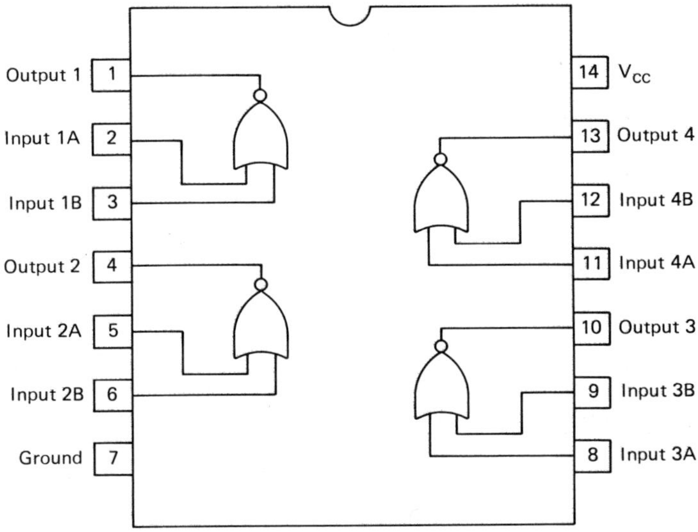

This device contains four independent NOR logic gates. Each gate consumes less than 2 μA of current while in the quiescent mode. Input current is typically less than 1 μA. The propagation delay increases as the supply voltage increases; it is 90 ns at +2 V and 18 ns at +5 V. Input capacitance is 10 pF. The truth table for each gate is:

INPUTS		OUTPUT
L	L	H
L	H	L
H	L	L
H	H	L

3 74HC03 Quad 2-Input NAND Gate with Open Drain Output

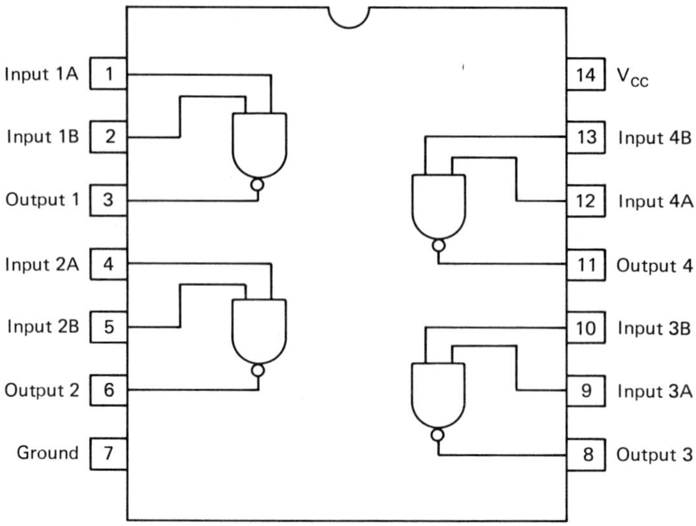

This device is very similar to the 74HC00 quad NAND gate, but each output has a high-performance n-channel MOS transistor. With a suitable pull-up resistor, this device's outputs are suitable for wired–AND purposes, as a LED driver, or for other applications that require a sinking current. Input current is typically less than 1 μA. Current consumption per gate in the quiescent mode is typically 2 μA. Propagation delay per gate is substantially higher than conventional HCMOS devices and decreases with increases in the supply voltage; it is 125 ns at +2 V and 25 ns at +5 V. The logic table is identical to that of the 74HC00.

4 74HC04 Hex Inverter
74HCT04 Hex Inverter with TTL–Compatible Inputs
74HCU04 Hex Unbuffered Inverter

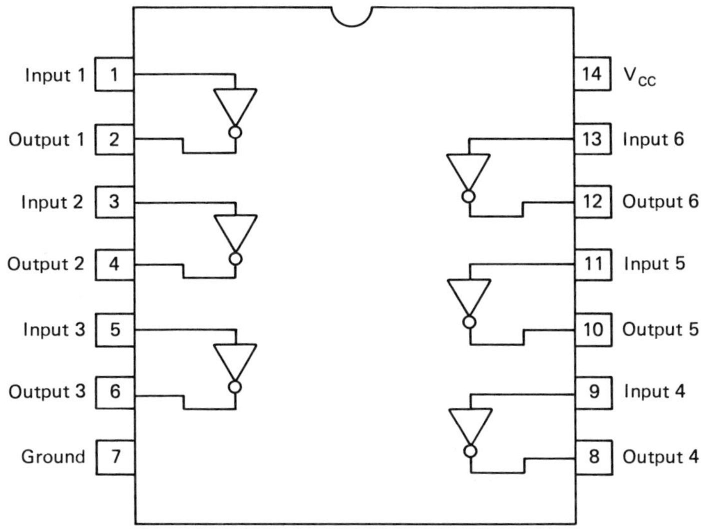

Each of these devices shares a common package and pin connections and consists of six independent single-stage inverter gates. The logic table for each inverter is:

INPUT	OUTPUT
L	H
H	L

Each device has a nominal input current of 1 μA and a quiescent current consumption of 2 μA per gate. The 74HC04 and 74HCT04 have propagation delays per gate of 20 ns, while the 74HCU04 is rated at 16 ns per gate. The 74HC04 is a standard HCMOS device with CMOS inputs and outputs compatible with CMOS, TTL, or NMOS. The 74HCT04 has inputs that are compatible with TTL and outputs that can directly drive CMOS, TTL, or NMOS; it is recommended for logic-level conversion applications. The 74HCU04 is suitable for a variety of linear applications, such as oscillators, for which the other two devices cannot be used.

5 74HC05 Hex Inverter with Open Drain Outputs

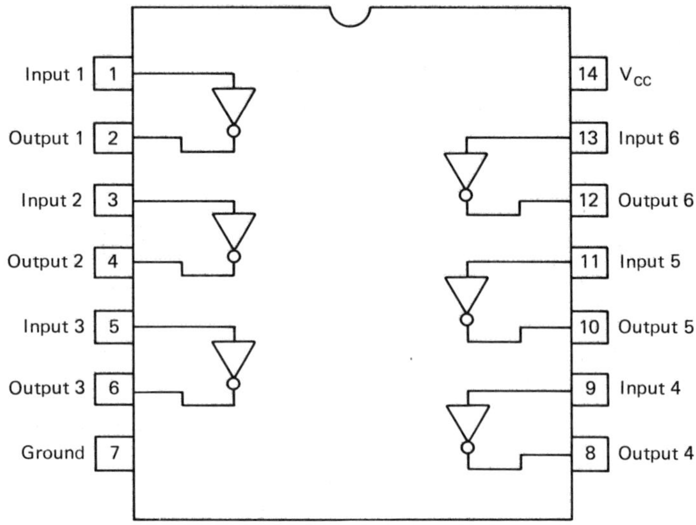

This device is very similar to the 74HC/HCT/HCU/04 family, but each of the outputs is fabricated using a high-performance n-channel MOS transistor. With a suitable pull-up resistor, each gate can be used in wired–AND or other applications requiring a sinking current. Input current is 1 μA, and power consumption per gate in the quiescent mode is 2 μA. Propagation delay is 100 ns at +2 V and 20 ns at +5 V. This device is also available in an HCT version. The logic table is the same as for 74HC/HCT/HCU/04 devices.

6 74HC08 Quad 2-Input AND Gate

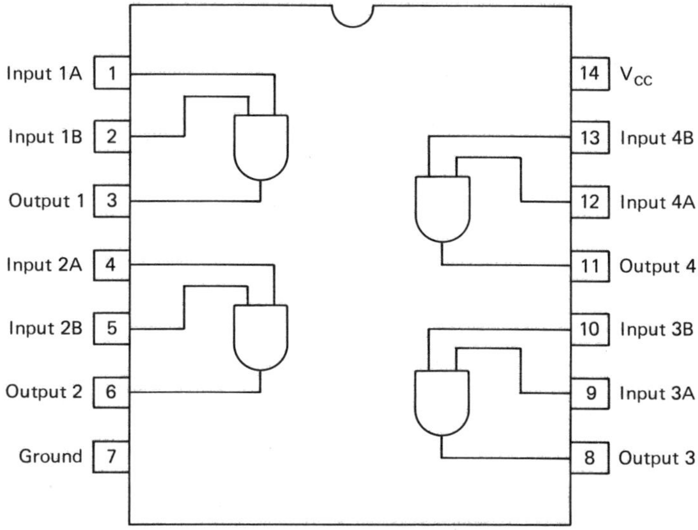

This device consists of four independent AND logic gates. Each gate consumes about 2 μA of current in the quiescent mode, and the input current is 1 μA. Propagation delay decreases with supply voltage; typical figures are 120 ns at +2 V and 24 ns at +5 V. Input capacitance is typically 10 pF. The logic table for each gate is:

INPUTS		OUTPUT
L	L	L
L	H	L
H	L	L
H	H	H

7 74HC10 Triple 3-Input NAND Gate

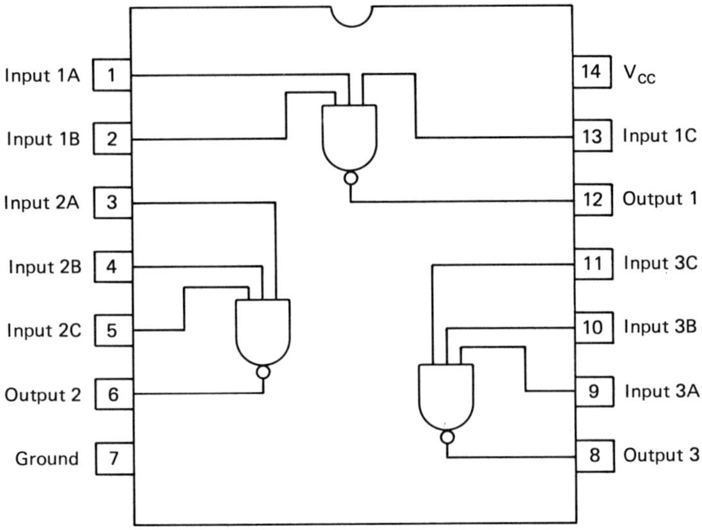

This device contains three separate 3-input NAND logic gates. Input current is 1 μA, and quiescent current consumption per gate is 2 μA. Propagation delay per gate decreases as the supply voltage increases, and it is rated at 95 ns at +2 V and 19 ns at +5 V. Input capacitance is 10 pF. The logic table for each gate is:

INPUTS			OUTPUT
L	X	X	H
X	L	X	H
X	X	L	H
H	H	H	L

8 74HC11 Triple 3-Input AND Gate

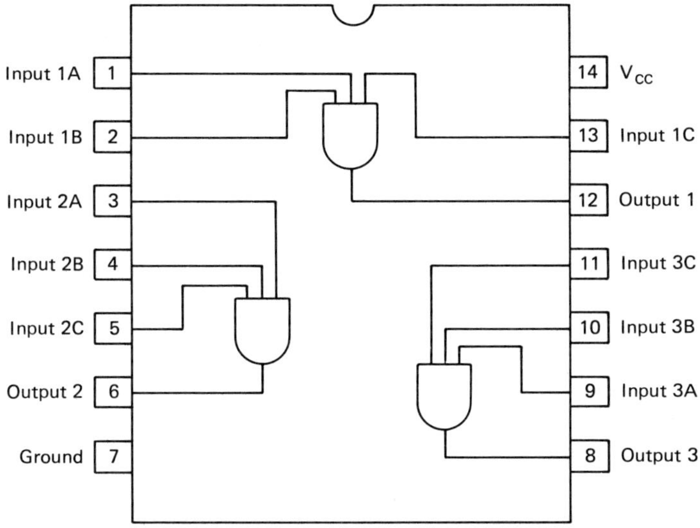

This device contains three separate 3-input AND logic gates. Input current is 1 μA, and quiescent current consumption is 2 μA. Propagation delay per gate decreases as the supply voltage increases; it is 125 ns at +2 V and 25 ns at +5 V. Input capacitance is 10 pF. The logic table for each gate is:

INPUTS			OUTPUT
L	X	X	L
X	L	X	L
X	X	L	L
H	H	H	H

9 74HC14 Hex Schmitt Trigger Inverter

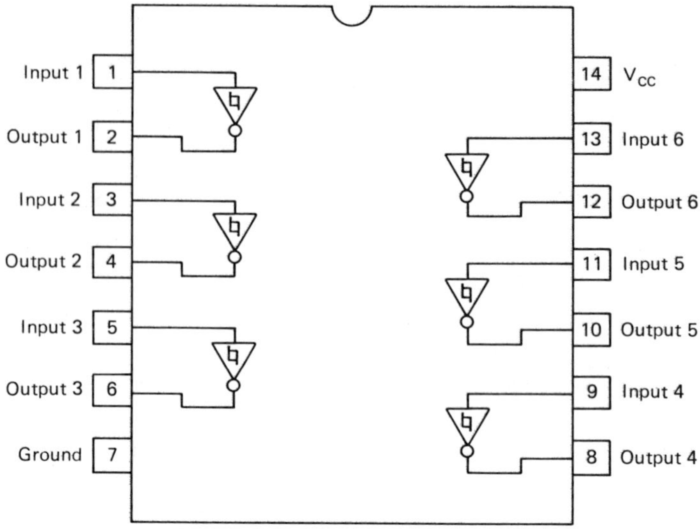

This device contains six independent Schmitt trigger inverters. These inverters function in much the same way as the inverters contained in the 74HC/HCT/HCU/04 device family except that they have a "snap action" suitable for squaring slowly rising or noisy inputs so they are "conditioned." Input current is 1 μA, and quiescent current consumption is 2 μA per gate. Propagation delay is 125 ns at +2 V and 25 ns at +5 V. The logic table for each gate is:

INPUTS	OUTPUT
L	H
H	L

10 74HC20 Dual 4-Input NAND Gate

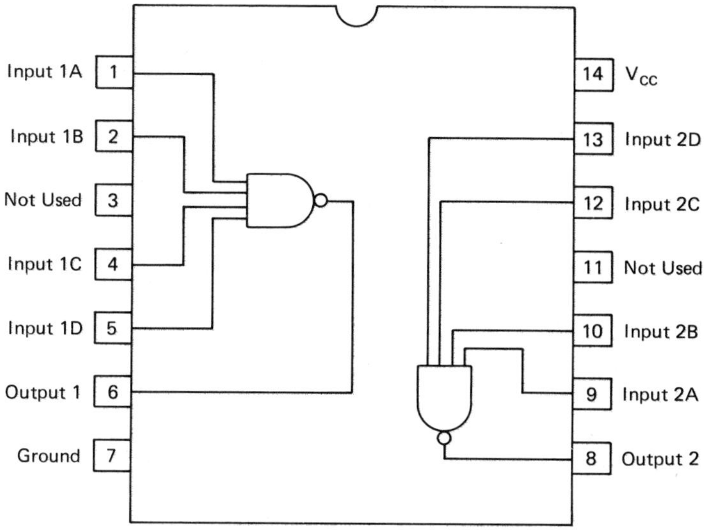

This device contains two separate 4-input NAND gates whose operation is similar to that of the 2-input NAND gates contained in the 74HC00 device. Input current is 1 μA, and current consumption in the quiescent mode is 2 μA. Input capacitance is 10 pF. Propagation delay is 90 ns at +2 V and 18 ns at +5 V. The logic table for each gate is:

INPUTS				OUTPUT
L	X	X	X	H
X	L	X	X	H
X	X	L	X	H
X	X	X	L	H
H	H	H	H	L

11 74HC21 Dual 4-Input AND Gate

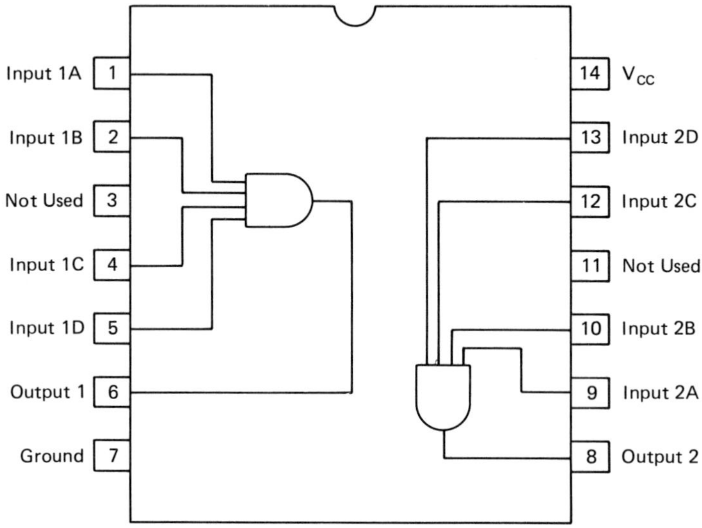

This device contains two independent AND gates whose operation is similar to those of the 74HC08. Input current is 1 μA, and quiescent mode current consumption per gate is 2 μA. Input capacitance is 10 pF. Propagation delay is 110 ns at +2 V and 22 ns at +5 V. The logic gate for each table is:

INPUTS				OUTPUT
L	X	X	X	L
X	L	X	X	L
X	X	L	X	L
X	X	X	L	L
H	H	H	H	H

12 74HC27 Triple 3-Input NOR Gate

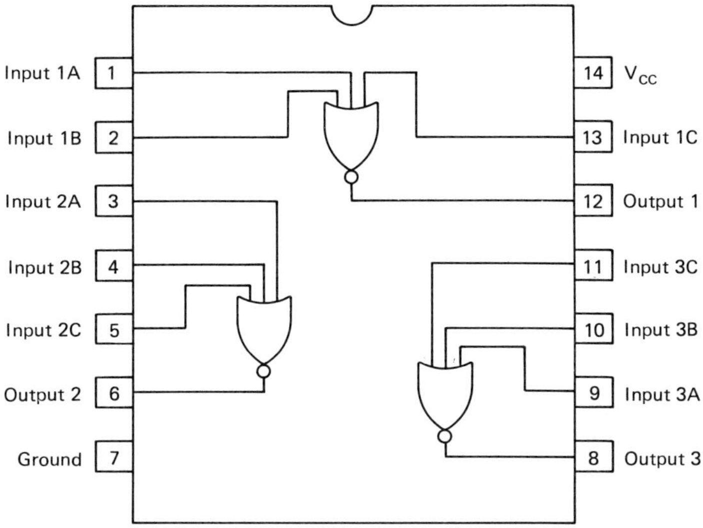

This device contains three independent NOR logic gates. Input current is 1 μA, and the maximum quiescent current consumption is 2 μA. Propagation delay per gate is 90 ns at +2 V supply voltage and 18 ns at +5 V. The logic table for each gate is:

INPUTS			OUTPUT
L	L	L	H
X	X	H	L
X	H	X	L
H	X	X	L

13 74HC30 8-Input NAND Gate

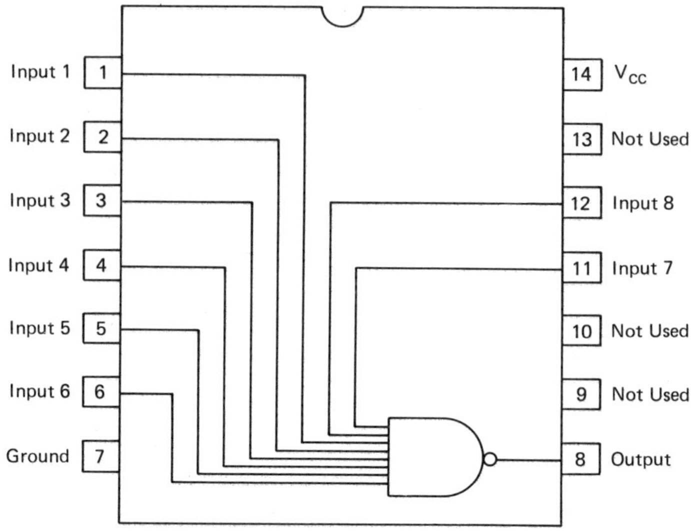

This is a single NAND gate with eight inputs. Input current is 1 μA, and quiescent current consumption is 2 μA. Propagation delay is 175 ns at +2 V and 35 ns at +5 V. The output of this device is low if all inputs are high; otherwise the output is high whenever one or more of the inputs are low.

14 74HC32 Quad 2-Input OR Gate

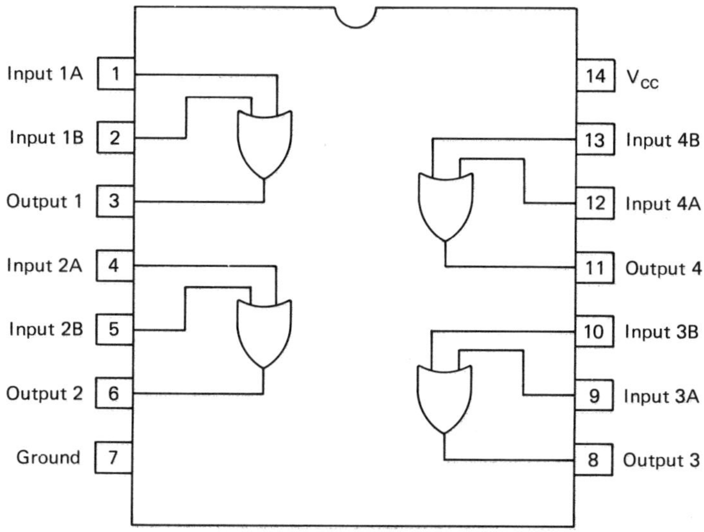

This device contains four separate OR logic gates. Input current is 1 μA, and quiescent mode current consumption is 2 μA. Propagation delay is 100 ns at +2 V and 20 ns at +5 V. Input capacitance is 10 pF. The logic table for this device is:

INPUTS		OUTPUT
L	L	L
L	H	H
H	L	H
H	H	H

15 74HC34 Hex Noninverting Buffer

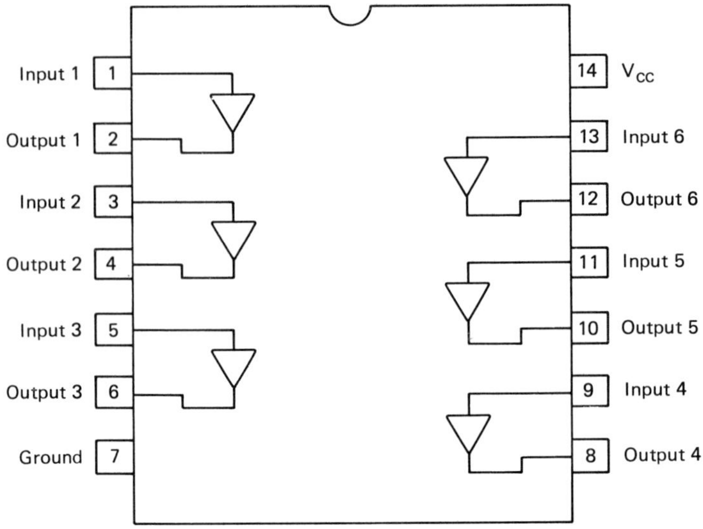

This device is similar to the 74HC/HCT/HCU/04 hex inverting buffer, but it is a noninverting type; the output is an "echo" of the input. Pin connections are identical to those of the hex inverting buffer device. Input current is 1 μA with a quiescent mode current consumption of 2 μA. Input capacitance is 10 pF. Propagation delay is 90 ns at +2 V and 18 ns at +5 V. This device is also available in HCT and open drain output versions. The logic table for this device is:

INPUT	OUTPUT
L	L
H	H

16 74HC42 1 of 10 Decoder

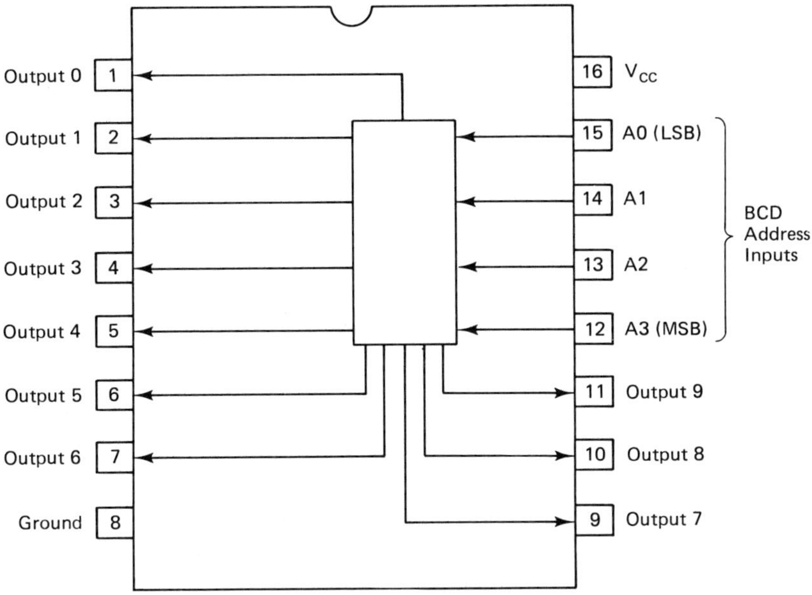

This device accepts a binary coded decimal (BCD) input and decodes it, activating one of ten active low outputs in response to the number represented by the BCD input. If the address inputs are given a signal with a hexadecimal equivalent of nine or greater, all outputs remain high. Pin 15 (A0) represents the least significant bit, and pin 12 (A3) represents the most significant bit of the BCD input. Input current is 1 μA, and quiescent mode current consumption is 8 μA. Propagation delay is 150 ns at +2 V and 30 ns at +5 V. The logic table for this device is:

INPUTS				OUTPUT GOING LOW
A3	A2	A1	A0	
L	L	L	L	Output 0
L	L	L	H	Output 1
L	L	H	L	Output 2
L	L	H	H	Output 3
L	H	L	L	Output 4
L	H	L	H	Output 5
L	H	H	L	Output 6
L	H	H	H	Output 7
H	L	L	L	Output 8
H	L	L	H	Output 9
H	L	H	L	All outputs high
H	L	H	H	All outputs high
H	H	L	L	All outputs high
H	H	L	H	All outputs high
H	H	H	L	All outputs high
H	H	H	H	All outputs high

17 74HC51 4-Input and 5-Input AND–NOR Gate

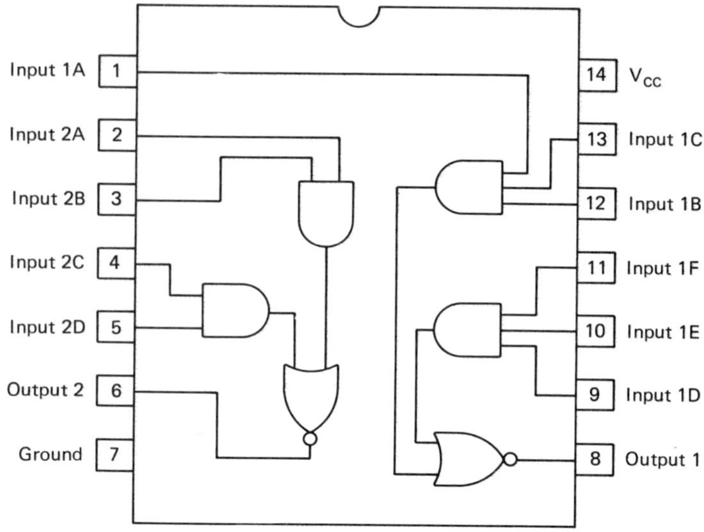

This device contains two independent logic networks, each composed of two AND gates and a NOR gate. One network uses two 3-input AND gates, while the other uses two 2-input AND gates. Input current is 1 μA, while quiescent current consumption by each network is 2 μA. Propagation delay is 125 ns at +2 V and 25 ns at +5 V. Input capacitance is 10 pF. Logic tables for each network are:

INPUTS (NETWORK #1)						OUTPUT 1
1A	1B	1C	1D	1E	1F	
H	H	H	X	X	X	L
X	X	X	H	H	H	L
All other combinations						H

INPUTS (NETWORK #2)				OUTPUT 2
2A	2B	2C	2D	
H	H	X	X	L
X	X	H	H	L
All other combinations				H

18 74HC58 4-Input and 5-Input AND–OR Gate

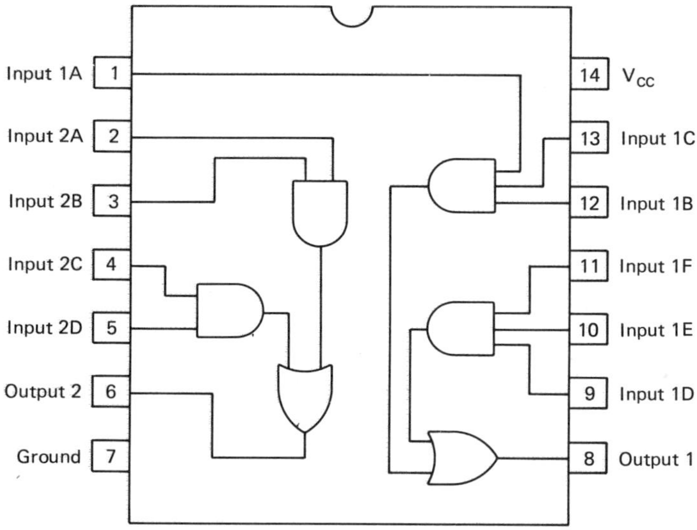

This device contains two independent logic networks, each consisting of two AND gates and a single OR gate. One network consists of two 3-input AND gates whose outputs are the inputs of an OR gate, while the other network consists of two 2-input AND gates whose outputs are the inputs of another OR gate. Current consumption per network in the quiescent mode is 2 μA, and input capacitance is 10 pF. Propagation delay is 125 ns at +2 V and 25 ns at +5 V. The logic tables for the two networks are:

INPUTS (NETWORK #1)						OUTPUT 1
1A	1B	1C	1D	1E	1F	
H	H	H	X	X	X	H
X	X	X	H	H	H	H
All other combinations						L

INPUTS (NETWORK #2)				OUTPUT 2
2A	2B	2C	2D	
H	H	X	X	H
X	X	H	H	H
All other combinations				L

19 74HC73 Dual J-K Flip-Flop with Reset

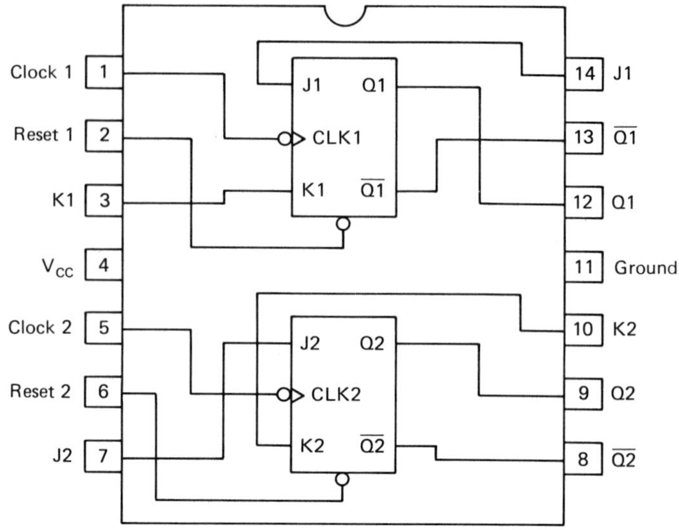

This device contains two independent J-K flip-flop circuits that have reset and clock inputs. Each flip-flop has complementary outputs, labeled Q and $\bar{Q}$. These devices are negative-edge triggered, meaning that the outputs change on the edge of a transition from a high to a low logic level in the clock signal. When the reset input is low, the flip-flop circuit is set to its initial state of Q being low and $\bar{Q}$ being high. To change from this state in response to changes in the J and K inputs or the clock signal, the reset input must be kept at a high level. Reset is asynchronous in operation and overrides the clock input signal. Each flip-flop is also capable of toggle operation, in which the outputs switch states in response to changes in the clock signal. The maximum clock frequency for operation is 6 MHz at +2 V and 30 MHz at +5 V. Propagation delay is 125 ns at +2 V and 25 ns at +5 V. Quiescent mode current consumption is 4 μA per flip-flop. The logic table for each flip-flop is:

INPUTS				OUTPUTS	
Reset	Clock	J	K	Q	$\bar{Q}$
L	X	X	X	L	H
H	F	L	L	No change	
H	F	L	H	L	H
H	F	H	L	H	L
H	F	H	H	Toggle	
H	L	X	X	No change	
H	H	X	X	No change	
H	R	X	X	No change	

20 74HC74 Dual D Flip-Flop with Set and Reset

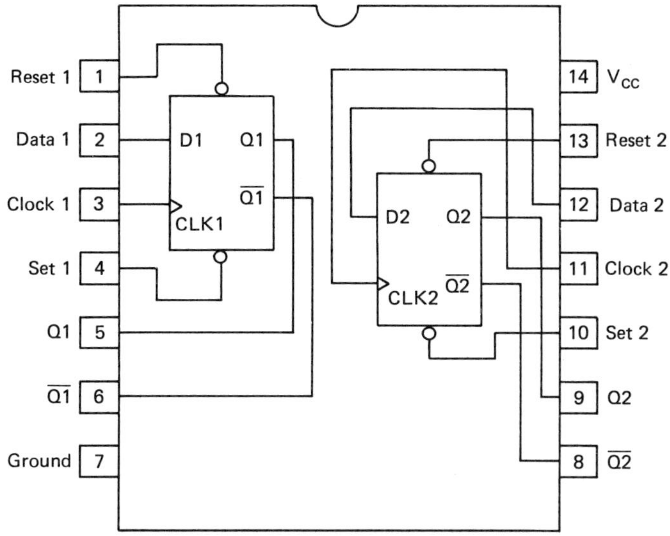

This device contains two independent D flip-flop circuits with reset and set inputs. Each circuit has a single date input labeled D and complementary outputs labeled Q and $\bar{Q}$. Each circuit is positive-edge triggered, meaning that the outputs can change states when the clock signal begins a transition from a low to a high logic level; this positive-edge transition causes the signal present at the D input to be transferred to the Q output, and $\bar{Q}$ then assumes the complementary logic level. The set and reset inputs are asynchronous and immediately affect the operation of the flip-flop regardless of the clock signal. Setting both reset and set to high will "freeze" both outputs at a high level. Maximum clock frequency is 6 MHz at +2 V and 30 MHz at +5 V. Propagation delay is 175 ns at +2 V and 35 ns at +5 V. Quiescent mode current consumption per flip-flop is 4 μA.

INPUTS				OUTPUTS	
Set	Reset	Clock	D	Q	$\bar{Q}$
L	H	X	X	H	L
H	L	X	X	L	H
L	L	X	X	H	H
H	H	R	H	H	L
H	H	R	L	L	H
H	H	L	X	No change	
H	H	H	X	No change	
H	H	F	X	No change	

21 74HC75 Dual 2-Input Transparent Latch

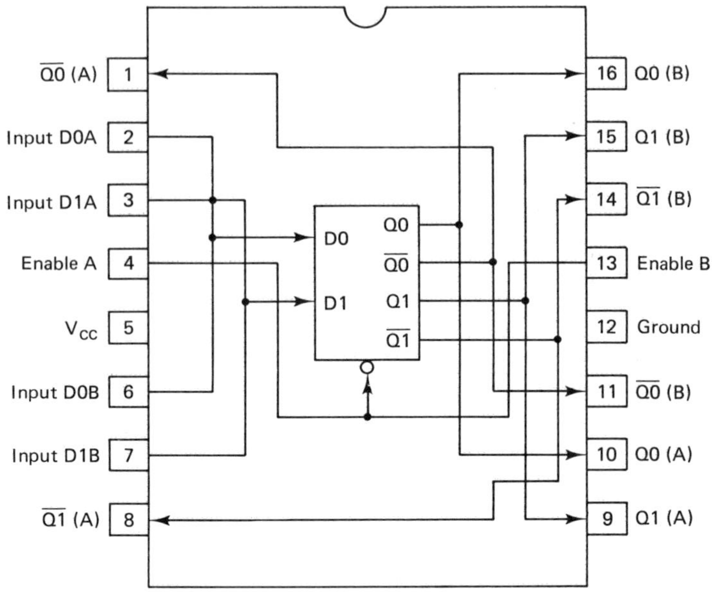

The 74HC75 can also be described as a *quad bistable transparent latch.* The internal circuitry of this device consists of four separate D flip-flop circuits grouped into two pairs controlled by two enable signals (pins 4 and 13). This in effect creates two two-bit memory storage registers, or latches. These latches are known as *transparent* because the outputs follow the inputs when the appropriate enable signal is high; when the enable signal is low, the circuit is "latched" and the outputs do not follow any further changes in the input signals as long as the enable signal remains low. The inputs at pins 2 and 3 are controlled by the enable input at pin 13, while the inputs at pins 6 and 7 are controlled by the enable signal at pin 4. *Note that the* Vcc *and ground pins are not in their usual locations.* The logic table for each latch is:

INPUTS		OUTPUTS	
D	Enable	Q	$\bar{Q}$
L	H	L	H
H	H	H	L
X	L	Latched	

22 74HC76 Dual J-K Flip-Flop with Set and Reset

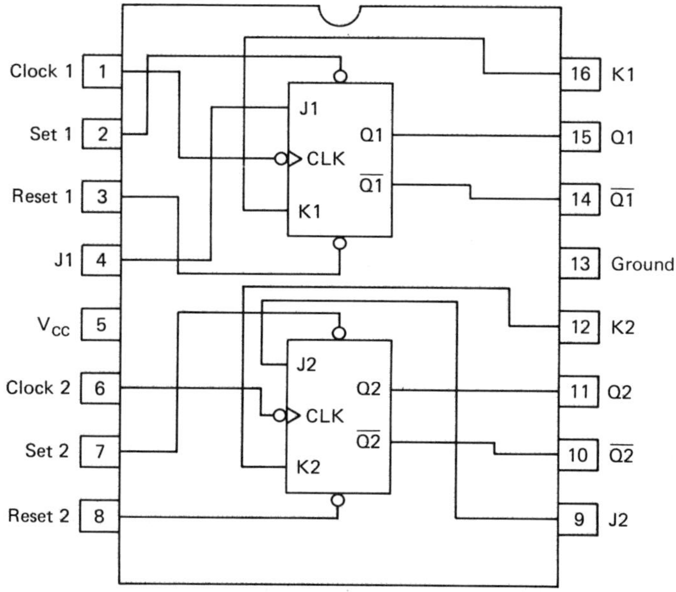

This device consists of two independent J-K flip-flop circuits with set and reset and complementary outputs. Each flip-flop is negative-edge triggered, while the set and reset inputs are asynchronous and override any clock signal change. Both set and reset inputs must be held high for normal operation. If both set and reset are high, and the J and K inputs are high, the outputs will toggle in response to the clock input. The maximum clock frequency is 6 MHz at +2 V and 30 MHz at +5 V. Maximum output transition time is 75 ns at +2 V and 15 ns at +5 V. *Note that the V_{cc} and ground pins are not in their usual locations.* The logic table for each flip-flop is:

INPUTS					OUTPUTS	
Set	Reset	Clock	J	K	Q	$\bar{Q}$
L	H	X	X	X	H	L
H	L	X	X	X	L	H
L	L	X	X	X	L	L
H	H	F	L	L	No change	
H	H	F	L	H	L	H
H	H	F	H	L	H	L
H	H	F	H	H	Toggle	
H	H	L	X	X	No change	
H	H	H	X	X	No change	
H	H	R	X	X	No change	

23 74HC85 4-Bit Magnitude Comparator

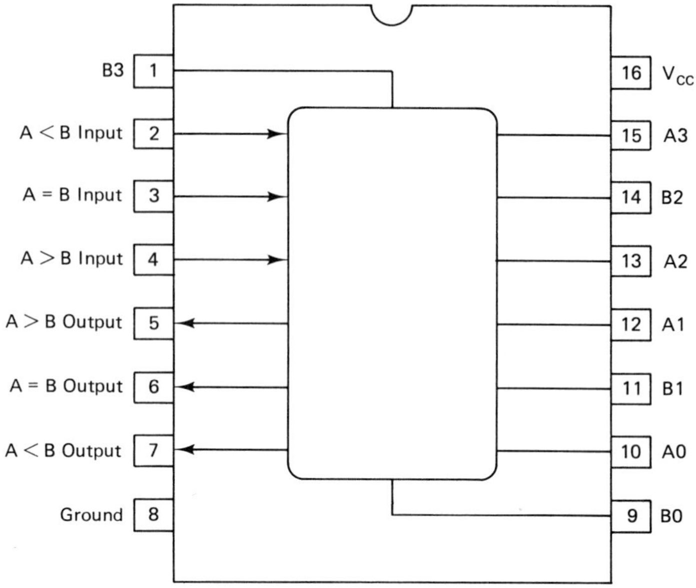

This device accepts two four-bit ("nibble") words, whose inputs are identified as A0 to A3 and B0 to B3, compares them, and produces a high-level output at the appropriate output and low-level outputs at the other two. This device also includes cascading inputs at pins 2, 3, and 4 so that devices can be cascaded to compare words larger than four bits; these cascading inputs are connected to the corresponding output on the preceding device. When a single device is used, pin 3 is held to a high level while pins 2 and 4 are held to a low level. Current consumption in the quiescent mode is 8 μA. Maximum output transition time is 75 ns at +2 V and 15 ns at +5 V.

24 74HC86 Quad 2-Input Exclusive OR Gate

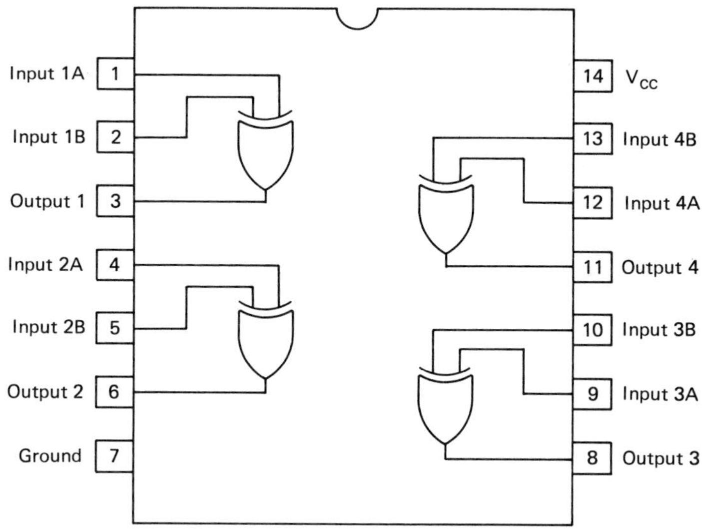

This device contains four independent exclusive OR logic gates. Propagation delay is 120 ns at +2 V and 24 ns at +5 V. Current consumption per gate in the quiescent mode is 2 μA, and input current is typically 1 μA. Input capacitance is 10 pF or less. The logic table for this device is:

INPUTS		OUTPUT
L	L	L
L	H	H
H	L	H
H	H	L

25 74HC93 4-Bit Binary Ripple Counter

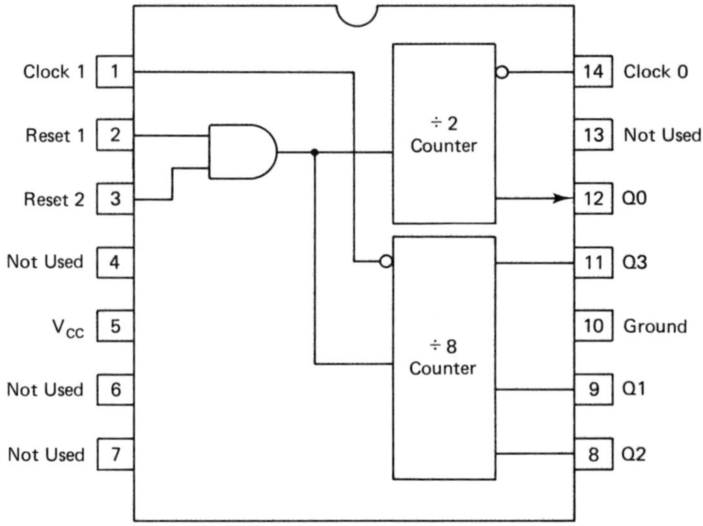

This device contains a divide-by-2 and divide-by-8 counter in a single package to produce a binary divide-by-16 counter. Each internal counter has a separate clock signal input, and both are negative-edge triggered. Setting both reset inputs (pins 2 and 3) to a high level sets all outputs (Q0 to Q3) to a low level. Each counter may be reset individually, and for normal operation both reset inputs should be held to a low level. This device can be configured for several different counting intervals. Maximum clock frequency is 100 MHz, and propagation delay is 12 ns at +5 V. This device is available in HCT as well. *Note the unusual positions of the* V_{cc} *and ground plans.*

26 74HC107 Dual J-K Flip-Flop with Reset

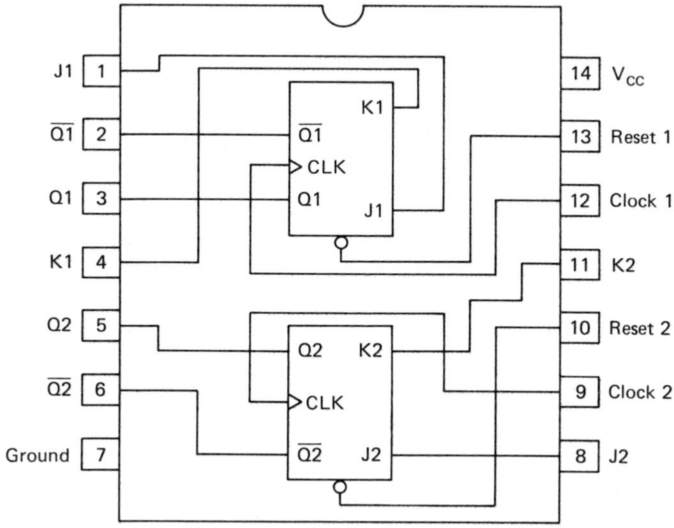

This device is identical in function and specifications to the 74HC73 but has a different pin connection arrangement. Consult the description of the 74HC73 for operational details and a logic table for the 74HC107.

27 74HC109 Dual J-K̄ Flip-Flop with Set and Reset

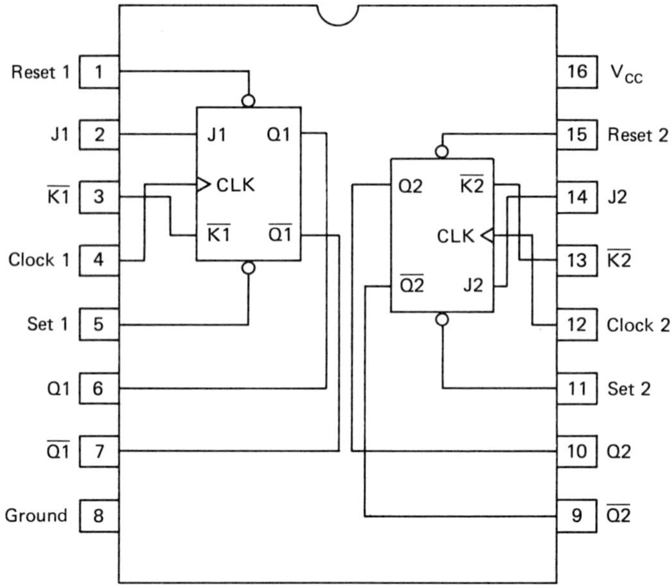

This device contains two independent J-K̄ flip-flops with individual set, reset, and clock inputs as well as complementary outputs. Each flip-flop is positive-edge triggered, and both the set and reset inputs must be held to a high level for normal operation. Set and reset inputs are asynchronous and override the J and K̄ inputs as well as the clock signal. Maximum clock frequency is 6 MHz at +2 V and 30 MHz at +5 V. Quiescent current consumption per flip-flop is 4 μA. The logic table for each flip-flop is:

INPUTS					OUTPUTS	
Set	Reset	Clock	J	K̄	Q	Q̄
L	H	X	X	X	H	L
H	L	X	X	X	L	H
L	L	X	X	X	H	H
H	H	R	L	L	L	H
H	H	R	H	L	Toggle	
H	H	R	L	H	No change	
H	H	R	H	H	H	L
H	H	L	X	X	No change	

28 74HC112 Dual J-K Flip-Flop with Set and Reset

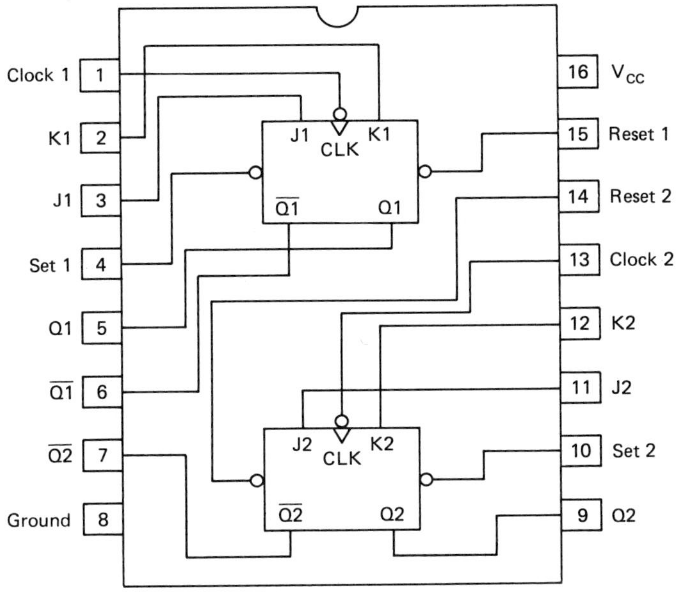

This device is identical in function and specifications to the 74HC76 but has a different pin connection arrangement. Consult the description of the 74HC76 for operational details and a logic table for the 74HC112.

29 74HC113 Dual J-K Flip Flop with Set

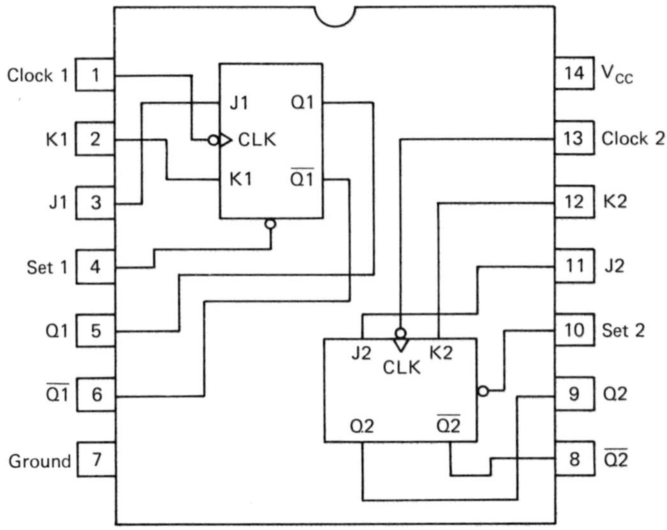

This device contains two independent J-K flip-flop circuits with set inputs and complementary outputs. Each flip-flop is negative-edge triggered, and the set input must be held to a high logic level for normal operation. The set inputs are asynchronous and override the clock, J, and K inputs. Maximum clock frequency is 6 MHz at 2 V and 30 MHz at 5 V; maximum quiescent current consumption is 4 μA. The logic table for each flip-flop is:

INPUTS				OUTPUTS	
Set	Clock	J	K	Q	$\bar{Q}$
L	X	X	X	H	L
H	F	L	L	No change	
H	F	L	H	L	H
H	F	H	L	H	L
H	F	H	H	Toggle	
H	H	X	X	No change	
H	L	X	X	No change	
H	R	X	X	No change	

30 74HC123 Dual Retriggerable Monostable Multivibrators with Resets

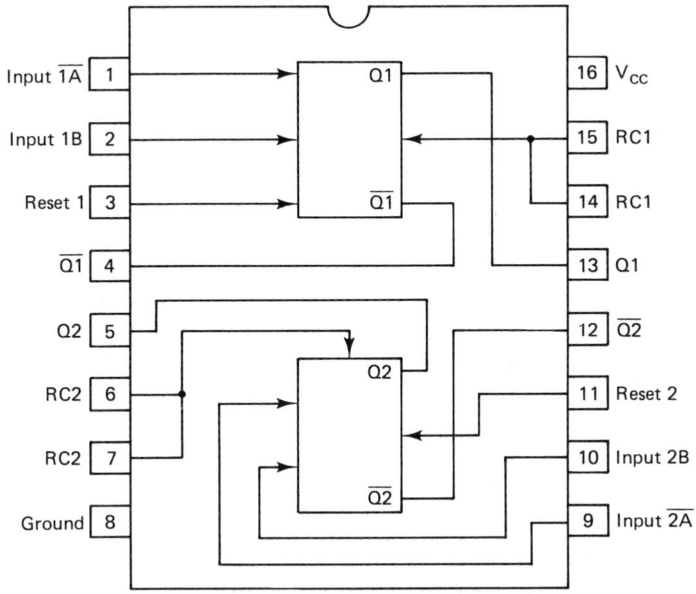

This device contains two independent monostable multivibrators, which each produce an output pulse whose duration is determined by timing resistors and capacitors connected to pins 14 and 15 or pins 6 and 7. The output pulse can be triggered in two ways: by holding input 1A or 2A low and bringing the corresponding B input from a low to a high level, or by holding input 1B or 2B high and bringing the corresponding A input from high to low. The reset input must be kept high for normal operation. The circuit may be retriggered to extend the length of the output pulse. Output times of both multivibrators when identical component values are used are accurate to within 2%. Quiescent current consumption at 5 V is 8 μA. The logic table for each multivibrator is:

INPUTS			OUTPUTS	
A	B	Reset	Q	$\bar{Q}$
H	X	H	L	H
X	L	H	L	H
L	R	H	HP	LP
F	H	H	HP	LP
X	X	L	L	H
L	H	R	HP	LP

31 74HC125 Quad 3-State Noninverting Buffers with Inverting Enable

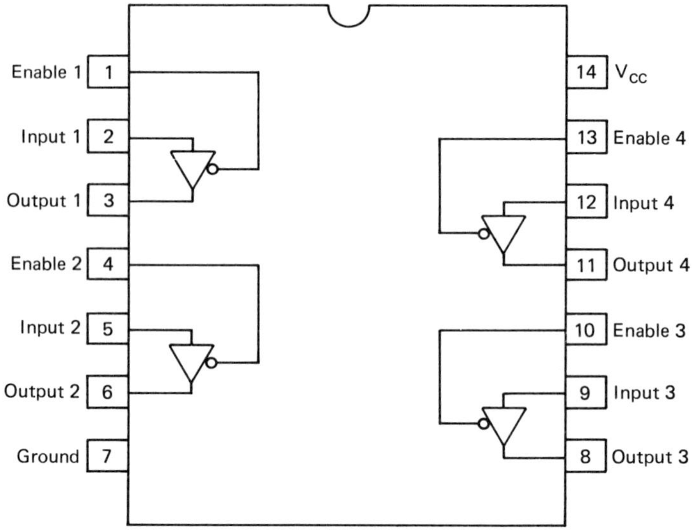

This device contains four separate buffer gates that serve to isolate logic signals from each other. The output will follow the input so long as the corresponding enable input is low. If the enable input is high, the output impedance will be very high. The function table for each buffer is:

INPUTS		OUTPUT
Input	Enable	
H	L	H
L	L	L
X	H	Z

32 74HC126 Quad 3-State Noninverting Buffers with Noninverting Enable

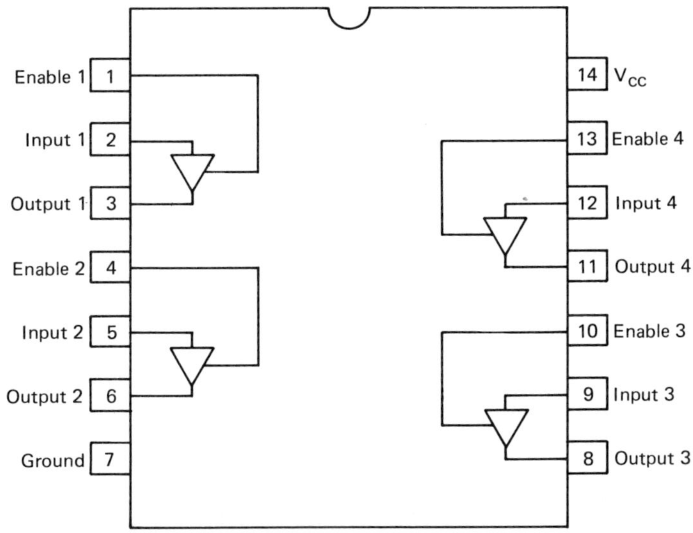

Very similar to the 74HC125, this device consists of four separate buffer gates whose output is identical to their input. Unlike the 74HC125, however, the enable signal for this device must be high. The function table for each buffer is:

INPUTS		OUTPUT
Input	Enable	
H	H	H
L	H	L
X	L	Z

33 74HC132 Quad 2-Input NAND Gate with Schmitt Trigger Inputs

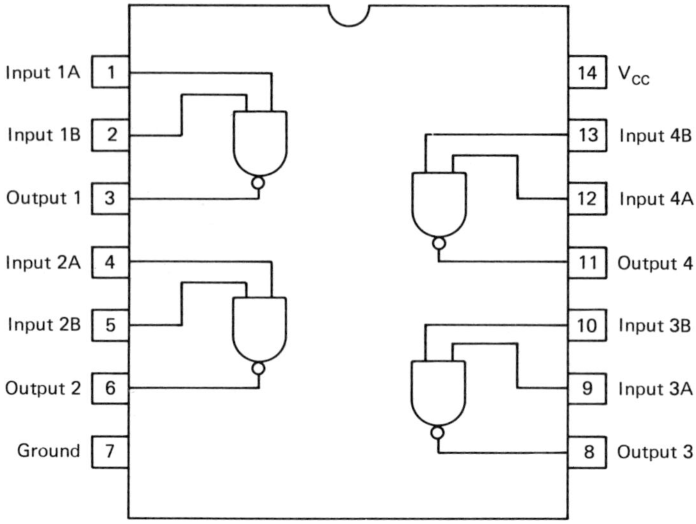

The logic table and pin connections for this device are identical to those of the 74HC00, but Schmitt triggers are incorporated at each input to enhance noise immunity and to "square up" slowly changing waveforms. It is used with irregular, "dirty" input waveforms or in situations where transient power line spikes and "noise" are a problem.

34 74HC133 13-Input NAND Gate

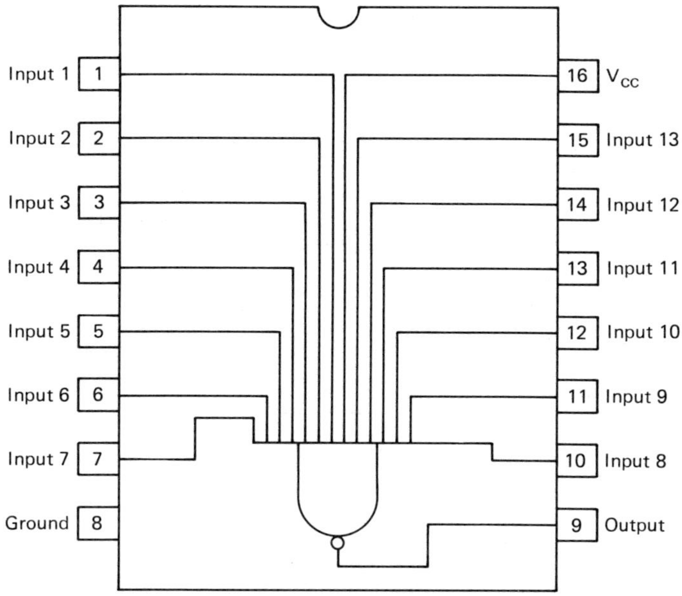

This device is a single NAND gate with 13 inputs. If all the inputs are high, the single output (pin 9) will be low. Any other combination of inputs will produce a high output. Maximum propagation delay is 35 ns at 5 V. Quiescent mode current consumption is 2 μA at 5 V.

35 74HC137 1 of 8 Decoder/Demultiplexer with Address Latch

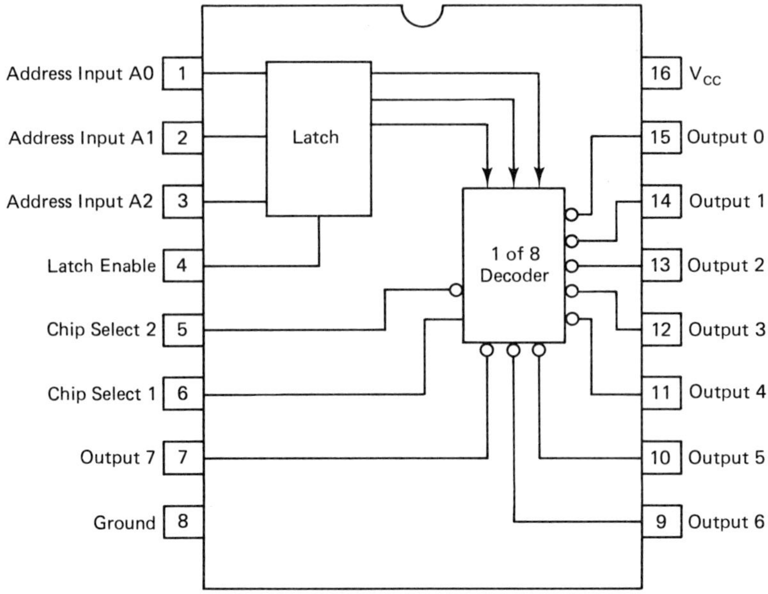

This device decodes a three-bit address input at pins 1 (A0), 2 (A1), and 3 (A2) and uses it to select one of the eight outputs. The device also has an internal transparent latch for storage of the address inputs. Two chip select inputs (pins 5 and 6) are provided to facilitate the demultiplexing and cascading functions; chip select 1 (CS1) should be kept high and chip select 2 (CS2) at low for normal operation. Keeping the chip select inputs at these levels allows the outputs to follow the address inputs. There is also a latch enable (LE) input (pin 4), which must be kept low for normal operation. Maximum quiescent mode current consumption is 8 μA. The logic table for this device is:

INPUTS						OUTPUTS							
LE	CS1	CS2	A0	A1	A2	0	1	2	3	4	5	6	7
X	X	H	X	X	X	H	H	H	H	H	H	H	H
X	L	X	X	X	X	H	H	H	H	H	H	H	H
L	H	L	L	L	L	L	H	H	H	H	H	H	H
L	H	L	L	L	H	H	L	H	H	H	H	H	H
L	H	L	L	H	L	H	H	L	H	H	H	H	H
L	H	L	L	H	H	H	H	H	L	H	H	H	H
L	H	L	H	L	L	H	H	H	H	L	H	H	H
L	H	L	H	L	H	H	H	H	H	H	L	H	H
L	H	L	H	H	L	H	H	H	H	H	H	L	H
L	H	L	H	H	H	H	H	H	H	H	H	H	L
H	H	L	X	X	X	No change							

36 74HC138 1 of 8 Decoder/Demultiplexer

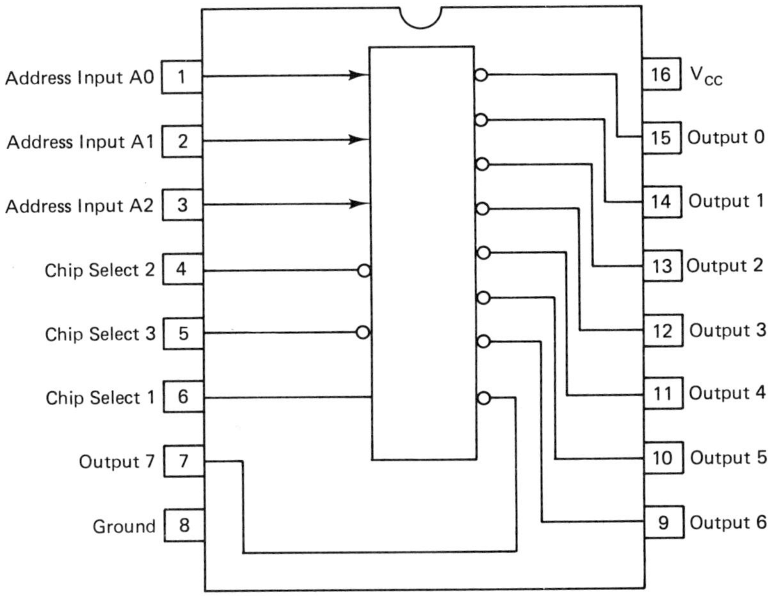

This device takes a three-bit address input at pins 1 (A0), 2 (A1), and 3 (A2) and uses it to select one of eight outputs. There are three chip select inputs. Chip select 1 (CS1) is normally held high, while chip select 2 (CS2) and chip select 3 (CS3) are normally held low. Maximum quiescent current consumption is 8 μA. The logic table for this device is:

INPUTS						OUTPUTS							
CS1	CS2	CS3	A2	A1	A0	0	1	2	3	4	5	6	7
X	X	H	X	X	X	H	H	H	H	H	H	H	H
X	H	X	X	X	X	H	H	H	H	H	H	H	H
L	X	X	X	X	X	H	H	H	H	H	H	H	H
H	L	L	L	L	L	L	H	H	H	H	H	H	H
H	L	L	L	L	H	H	L	H	H	H	H	H	H
H	L	L	L	H	L	H	H	L	H	H	H	H	H
H	L	L	L	H	H	H	H	H	L	H	H	H	H
H	L	L	H	L	L	H	H	H	H	L	H	H	H
H	L	L	H	L	H	H	H	H	H	H	L	H	H
H	L	L	H	H	L	H	H	H	H	H	H	L	H
H	L	L	H	H	H	H	H	H	H	H	H	H	L

37 74HC139 Dual 1 of 4 Decoder/Demultiplexer

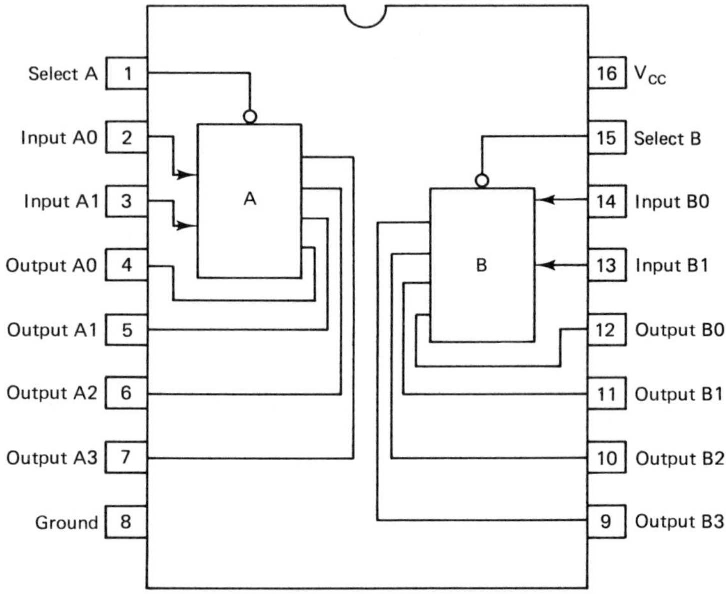

This device contains two independent one of four decoders, each of which decodes a two-bit address input (A0 and A1 or B0 and B1) and activates one of four outputs. Each decoder has a select input that is held low for normal operation. Each separate decoder/demultiplexer can be cascaded. Maximum quiescent mode current consumption is 8 μA per decoder. The logic table for each device is:

INPUTS			OUTPUTS			
Select	A1	A0	0	1	2	3
H	X	X	H	H	H	H
L	L	L	L	H	H	H
L	L	H	H	L	H	H
L	H	L	H	H	L	H
L	H	H	H	H	H	L

38 74HC147 Decimal to BCD Encoder

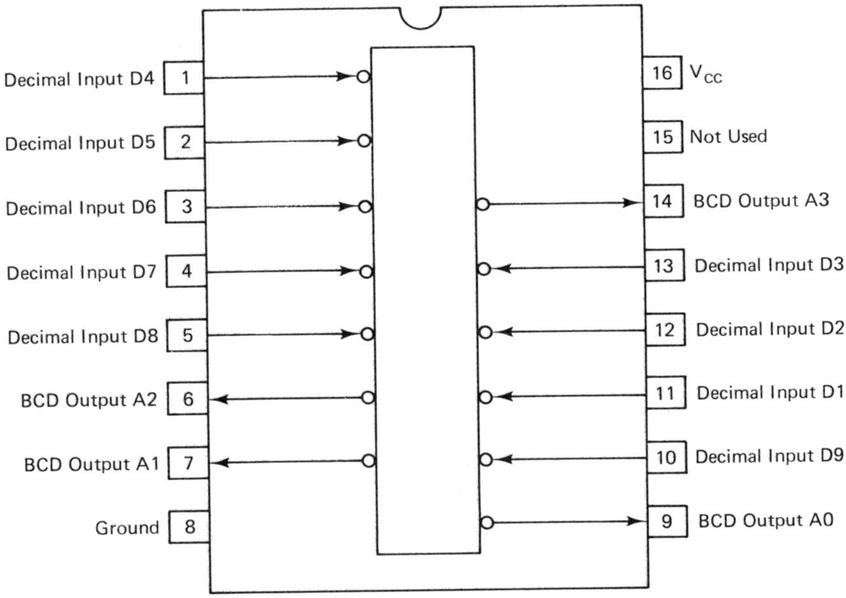

This device takes nine decimal data inputs (D1 through D9) and encodes them into a BCD output appearing at four outputs labeled A3 through A0. The implied decimal zero is represented when all data inputs are at a high level. The maximum propagation delay is 220 ns at 2 V and 44 ns at 5 V. The logic table for this device is:

INPUTS									OUTPUTS			
D9	D8	D7	D6	D5	D4	D3	D3	D1	A3	A2	A1	A0
H	H	H	H	H	H	H	H	H	H	H	H	H
H	H	H	H	H	H	H	H	L	H	H	H	L
H	H	H	H	H	H	H	L	X	H	H	L	H
H	H	H	H	H	H	L	X	X	H	H	L	L
H	H	H	H	H	L	X	X	X	H	L	H	H
H	H	H	H	L	X	X	X	X	H	L	H	L
H	H	H	L	X	X	X	X	X	H	L	L	H
H	H	L	X	X	X	X	X	X	H	L	L	L
H	L	X	X	X	X	X	X	X	L	H	H	H
L	X	X	X	X	X	X	X	X	L	H	H	L

39 74HC151 8-Input Data Selector/Multiplexer

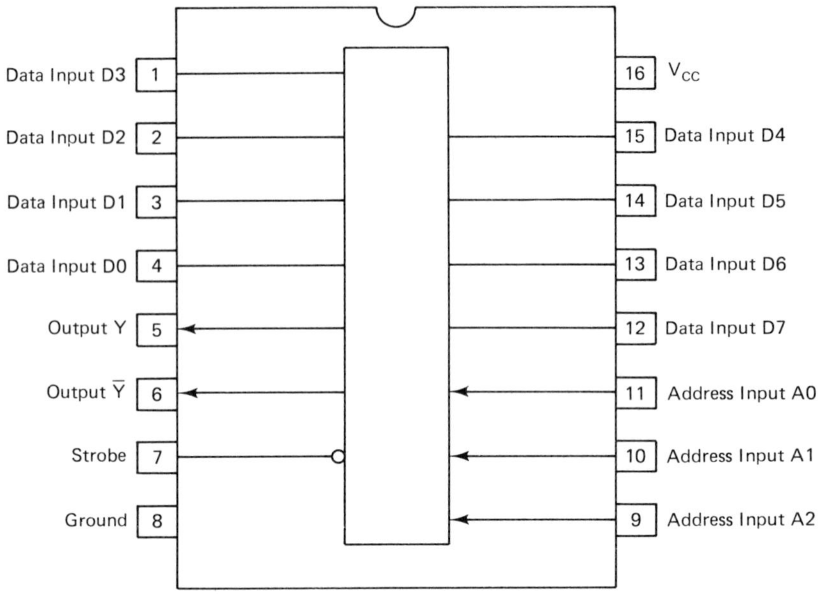

This device selects one of eight binary data inputs (labeled D0 through D7) as determined by the address inputs (A0, A1, and A2) and allows one of the bits to appear at Y and its complement at $\overline{Y}$. The strobe input must be held at a low level for normal operation; if it is high, the Y output is at a low logic level while Y is high. Propagation delay is 185 ns at 2 V and 37 ns at 5 V; maximum quiescent mode current consumption is 8 μA. The function table is:

INPUTS				OUTPUTS	
A2	A1	A0	Strobe	Y	$\overline{Y}$
X	X	X	H	L	H
L	L	L	L	D0	D0
L	L	H	L	D1	D1
L	H	L	L	D2	D2
L	H	H	L	D3	D3
H	L	L	L	D4	D4
H	L	H	L	D5	D5
H	H	L	L	D6	D6
H	H	H	L	D7	D7

40 74HC153 Dual 4-Input Data Selector/Multiplexer

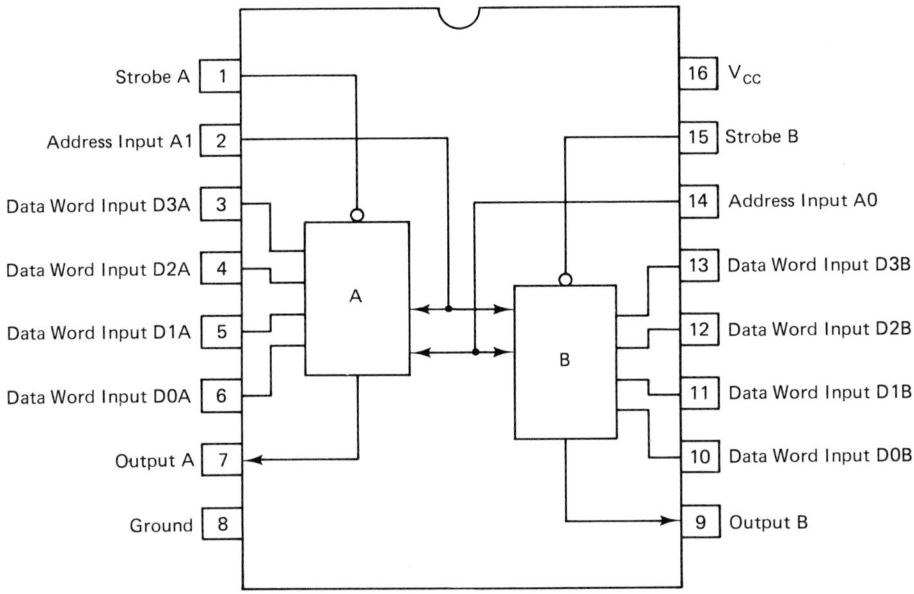

This device consists of two independent 4-input data selector/multiplexer circuits. The address inputs at pin 2 (A1) and pin 14 (A0) select one of the four data inputs (D0A to D3A or D0B to D3B) of each multiplexer and allow the input (high or low) to appear at the noninverting output of each section. Each section has a strobe input that must be kept low for normal operation; if it is high, the output of the section will be low. Propagation delay is 140 ns at 2 V and 28 ns at 5 V. The function table for each section is:

INPUTS			OUTPUT
A1	A0	Strobe	
X	X	H	L
L	L	L	D0
L	H	L	D1
H	L	L	D2
H	H	L	D3

41 74HC154 1 of 16 Decoder/Multiplexer

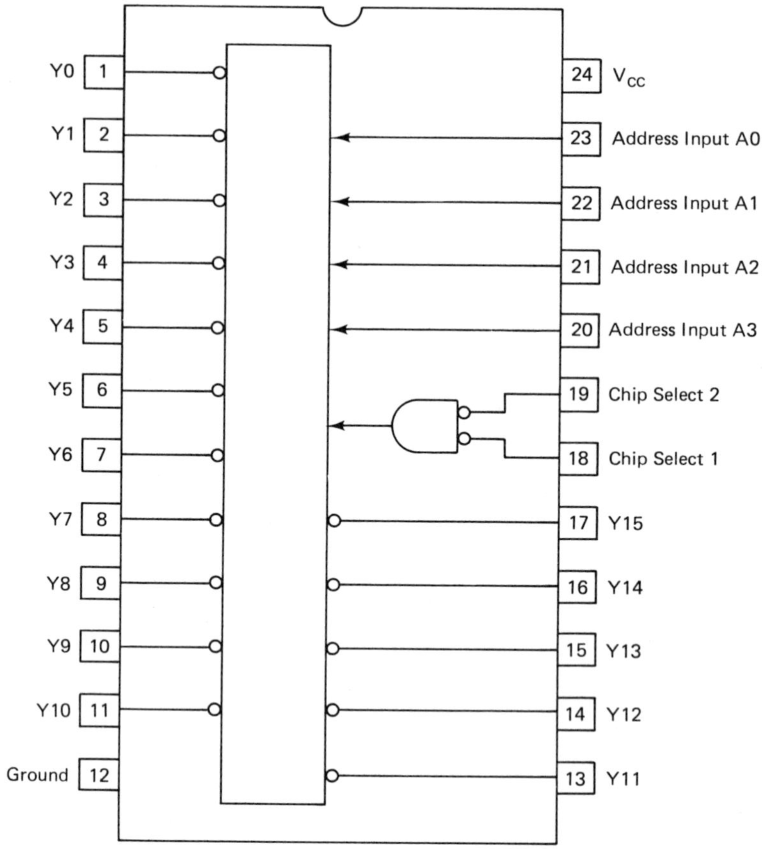

This device selects one of the 16 device outputs (labeled Y0 through Y15) according to logic levels applied to the four binary address inputs (labeled A0 to A3). For normal operation, both chip select inputs (CS1 and CS2) must be held low. If either or both chip select inputs are high, all outputs will be high. Maximum quiescent mode supply current is 8 μA. The propagation delay is 190 ns at 2 V and 38 ns at 5 V. The function table for this device is:

INPUTS				SELECTED OUTPUT
1	2	4	8	
0	0	0	0	0
0	0	0	1	1
0	0	1	0	2
0	0	1	1	3
0	1	0	0	4
0	1	0	1	5
0	1	1	0	6
0	1	1	1	7
1	0	0	0	8
1	0	0	1	9
1	0	1	0	10
1	0	1	1	11
1	1	0	0	12
1	1	0	1	13
1	1	1	0	14
1	1	1	1	15

42 74HC157 Quad 2-Input Data Selectors/Multiplexers with Noninverting Outputs

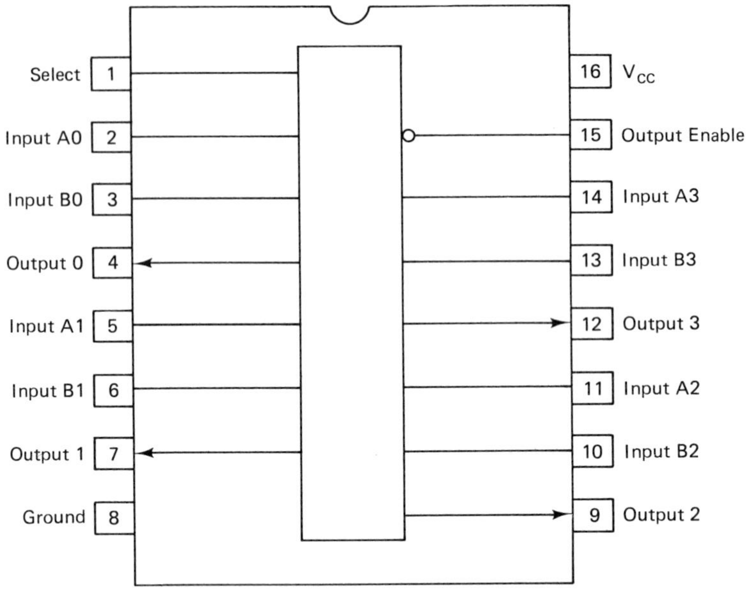

This device takes two four-bit "nibbles" at inputs A0 to A3 or B0 to B3 and routes them to the four outputs. The select input (pin 1) controls which nibble appears at the outputs. If the select input is low, inputs A0 to A3 appear at the outputs. If the select input is high, inputs B0 to B3 appear at the outputs. The device functions normally only when the output enable input (pin 15) is low; otherwise all outputs are low. Propagation delay is 125 ns at 2 V and 25 ns at 5 V. The function table for this device is:

INPUTS		OUTPUTS
Enable	Select	
H	X	All low
L	L	A0 to A3
L	H	B0 to B3

43 74HC158 Quad 2-Input Data Selectors/Multiplexers with Inverting Outputs

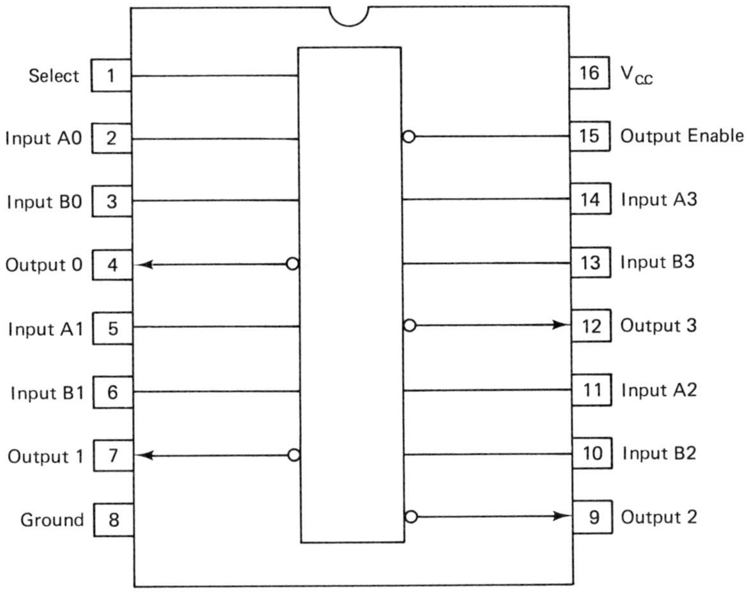

This device is similar in design and operation to the 74HC157, but the outputs are inverted from those appearing at the inputs. The function table for this device is:

INPUTS		OUTPUTS
Enable	Select	
H	X	All high
L	L	A0 to A3
L	H	B0 to B3

44 74HC160 BCD Presettable Counter with Asynchronous Reset

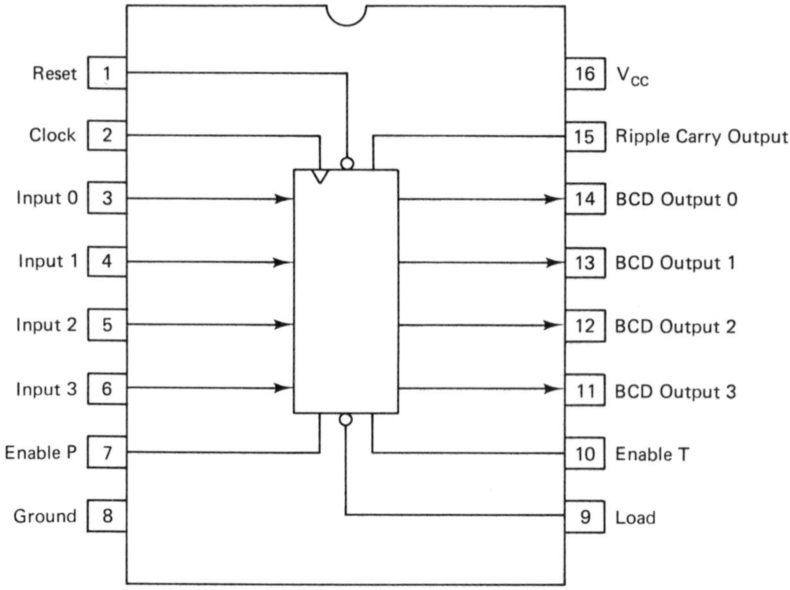

This device is a presettable (programmable) binary-coded decimal counter with four inputs (labeled 0 to 3) and four outputs (also labeled 0 to 3). Counting action is controlled by a positive-edge signal at pin 2, and the count advances by one with each clock pulse. The reset input (pin 1) is asynchronous; a low logic level signal to it will set all outputs to a low level regardless of the other inputs. It should be kept at a high logic level for normal operation. There are two enable inputs, labeled P (pin 7) and T (pin 10). Both must be at a high level for normal counting operation. A low-level signal at the load input (pin 9) disables the counting action and causes the data present at the inputs to be loaded into the counter on the next positive-edge clock signal. A low signal at the load inputs causes loading regardless of the signals at the enable P and enable T inputs. There is a ripple carry output (pin 15) to allow cascading of devices. Maximum clock frequency is 6 MHz at 2 V and 30 MHz at 5 V. Quiescent mode current consumption is 8 μA. The function table for this device is:

INPUTS					OUTPUTS
Clock	Reset	Load	P	T	
R	L	X	X	X	All low
R	H	L	X	X	Load input data
R	H	H	H	H	Count
R	H	H	L	X	No count
R	H	H	X	L	No count

45 74HC161 Binary Presettable Counter with Asynchronous Reset

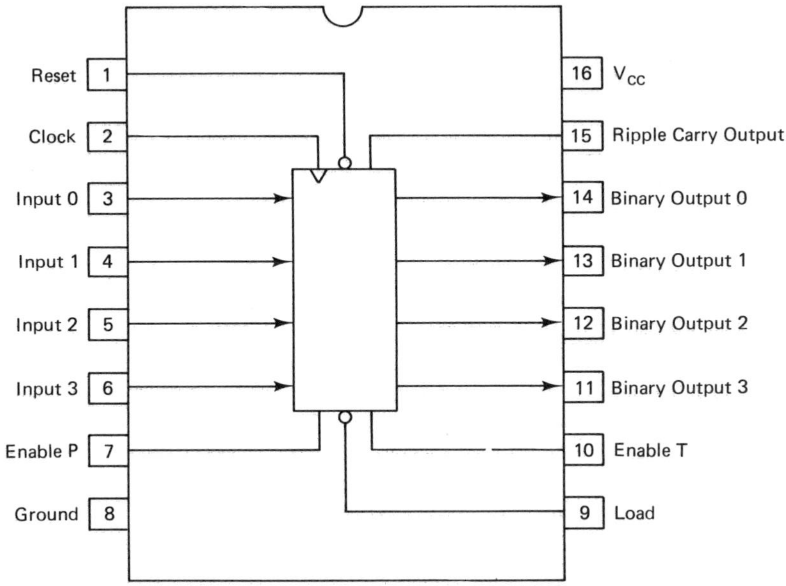

This device is a presettable (programmable) four-bit binary counter with asynchronous reset. Except for its binary counting, this device is very similar in operation to the 74HC160. Its inputs and outputs are identical to those of the 74HC160, and the various inputs accept the same signals and produce the same results. The function table is also the same, as are the clock frequency and current consumption figures. The only difference is that the 74HC160 is a BCD counter, whereas the 74HC161 is a binary counter.

46 74HC162 BCD Presettable Counter with Synchronous Reset

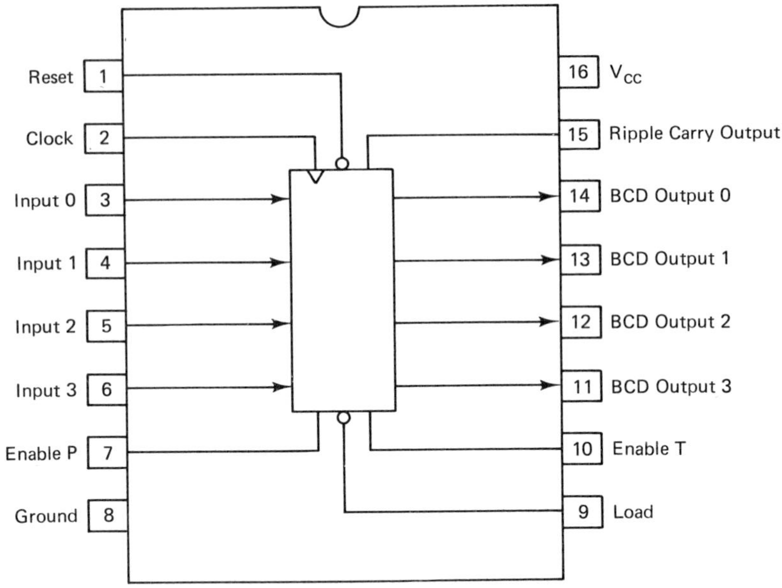

This device is a presettable (programmable) BCD counter that is very similar to the 74HC160 except that the reset input can function *only* on a positive edge of the clock signal. Its inputs and outputs are identical to those of the 74HC160, and the various inputs accept the same signals and produce the same results. The function table is also the same, as are the clock frequency and current consumption figures. The only difference is that the 74HC160 has an asynchronous reset (one that operates independently of the clock input), while the 74HC162 reset can only operate on a positive edge of a clock signal.

47 74HC163 Binary Presettable Counter with Synchronous Reset

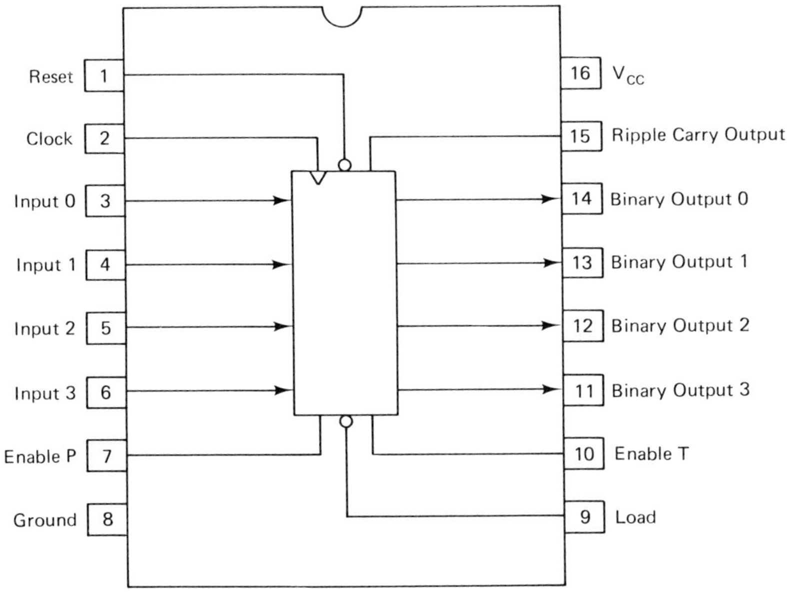

This device is a presettable programmable four-bit binary counter similar to the 74HC161 except that the reset input can function *only* on a positive edge of the clock signal. Its inputs and outputs are identical to those of the 74HC161 (and 74HC160), and the various inputs accept the same signals and produce the same results. The function table is the same as that for the 74HC160/74HC161/74HC162, as are the clock frequency and current consumption figures. The only difference is that the 74HC161 has an asynchronous reset (one that operates independently of the clock input), whereas the 74HC162 reset can only operate on a positive edge of a clock signal.

48 74HC164 8-Bit Serial Input/Parallel Output Shift Register

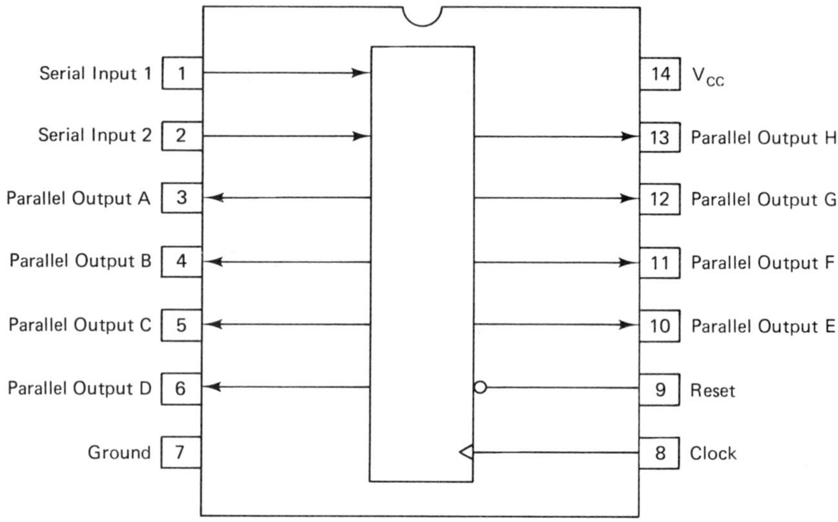

This device is a right-shift only shift register with two serial inputs (labeled 1 and 2) and eight parallel outputs (labeled A through H). One of the serial inputs can be used as an enable input. The 74HC164 uses positive-edge clocking, and data are entered on the positive edge of each clock pulse. The reset input is asynchronous and overrides the clock and serial inputs; it should be kept at a high level for normal operation. When reset is low, all the parallel outputs are set to a low level. The serial input to which the data are applied is known as the "D" input. The other serial input should be held at a high logic level. In normal operation, the "D" input appears at the "A" output (pin 3), and the remaining outputs have data shifted from the previous output stage (i.e., the signal appearing at the "B" output is shifted to the "C" output when a new clock pulse is received). Maximum clock frequency is 6 MHz at 2 V and 30 MHz at 5 V. In the following logic table, "D" represents the data input:

INPUTS				OUTPUTS	
Reset	Clock	1	2	A	B through H
L	X	X	X	L	All low
H	F	X	X	No change from previous stage	
H	R	H	D	D	Data shifted from previous stage
H	R	D	H	D	Data shifted from previous stage

49 74HC165 8-Bit Serial or Parallel Input/Serial Output Shift Register

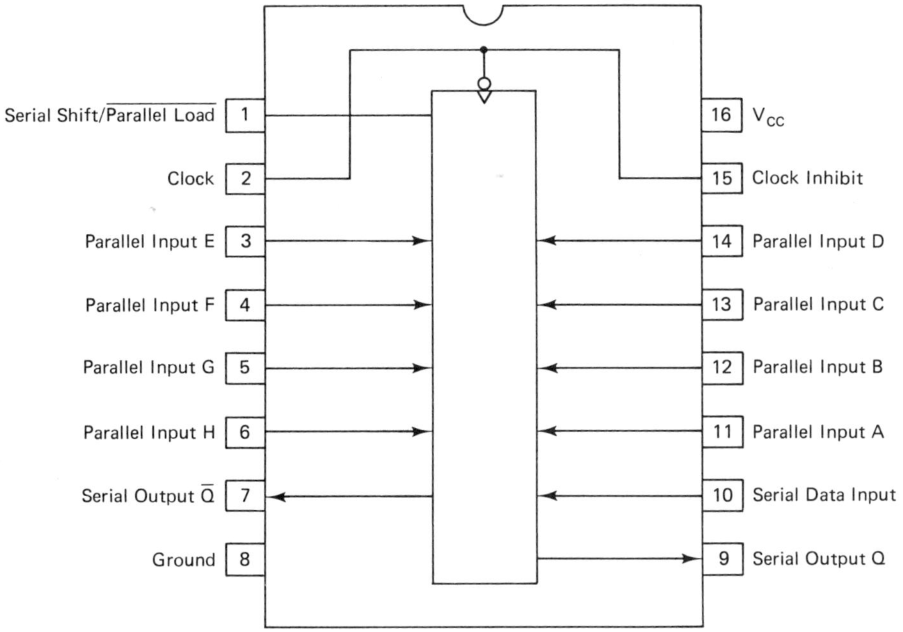

This device is a right-shift only shift register that can accept data in eight-bit parallel or serial form and output these data to complementary serial outputs. The serial shift/parallel load (S/P) input at pin 1 controls whether data are loaded in serial or parallel form. When pin 1 is low, data are loaded asynchronously in parallel at the inputs labeled A through H. The signal present at the A input appears at output Q (pin 9), and its complement appears at output $\bar{Q}$ (pin 7), followed by the signals present at inputs B, C, D, and so on. When pin 1 is high, data are loaded in serial form from the serial input (SI) at pin 10 on the positive edge of a clock pulse applied to the clock input (pin 2) or the clock inhibit (CI) input (pin 15). It is possible to combine two independent clock signals by simultaneously applying them to the clock and clock inhibit inputs, although this is seldom done. Maximum clock frequency is 6 MHz at 2 V and 30 MHz at 5 V. Maximum quiescent mode supply current is 8 μA. The function table is:

INPUTS				RESULT
S/P	Clock	CI	SI	
L	X	X	X	Asynchronous parallel load H
H	R	L	X	Serial shift via clock
H	L	R	X	Serial shift via clock inhibit
H	X	H	X	Inhibited clock
H	H	X	X	Inhibited clock
H	L	L	X	No clock

50 74HC166 8-Bit Parallel Input/Serial Output Shift Register

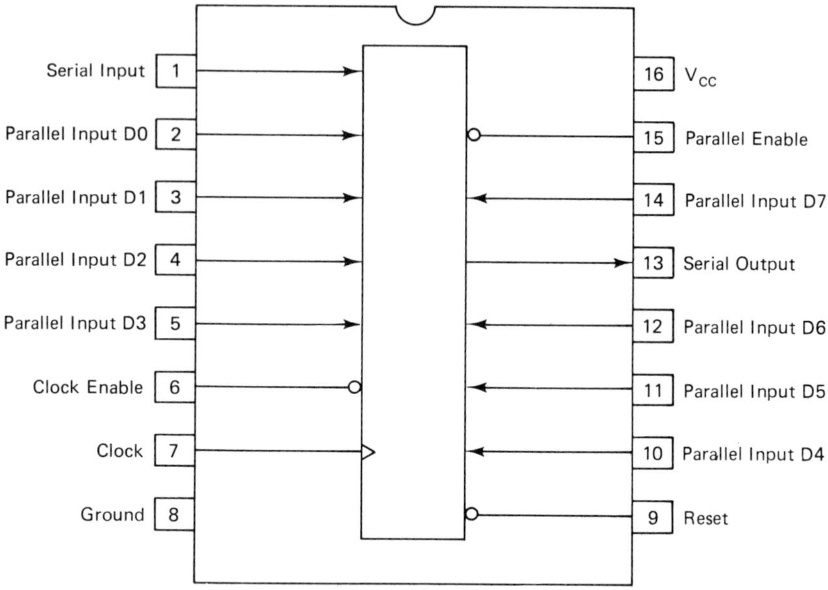

This device is a right-shift only shift register that accepts data in parallel form at eight inputs (labeled D0 through D7) and produces a serial output (SO) at pin 13. There is also a serial input (SI) at pin 1, which is primarily intended for use when cascading devices. Unlike the 74HC165, loading of the data present at the parallel input is synchronous and occurs only on positive-edge clock signals. The parallel enable (PE) input at pin 15 must be kept low to load parallel data. Either the clock input at pin 7 or the clock enable (CE) at pin 6 can be used as a clock signal input; the normal procedure is to hold the CE input low and apply the clock signal to the clock input at pin 7. Pin 9 is a reset input; a low signal to this input will force the output low and clear all internal registers. Maximum clock frequency is 19 MHz at 2 V and 57 MHz at 5 V. The function table for this device is:

INPUTS					OUTPUTS	RESULT
PE	CE	Clock	SI	D0–D7	SO	
L	L	R	X	All low	L	Parallel load
L	L	R	X	All high	H	Parallel load
H	L	R	L	X	X	Serial shifting
H	L	R	H	X	X	Serial shifting
X	H	X	X	X	X	Hold (do nothing)

51 74HC173 Quad 3-State D Flip-Flop with Common Clock and Reset

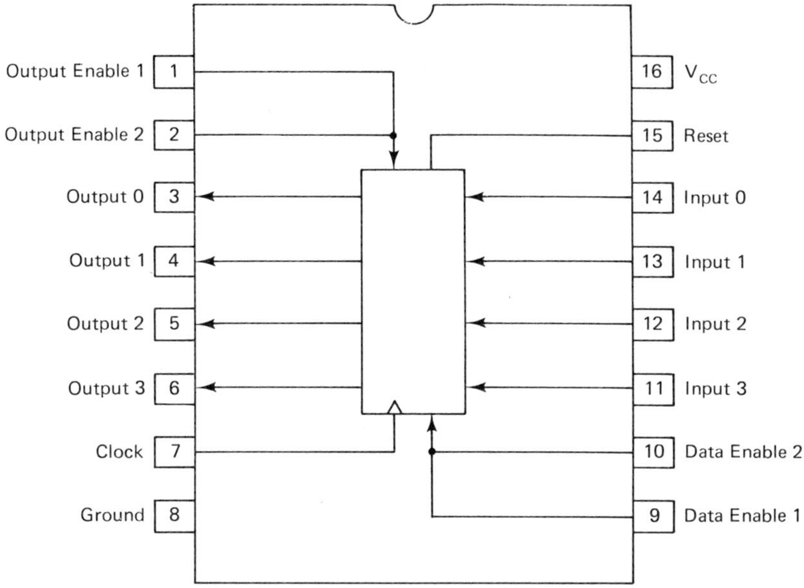

This device consists of four separate D flip-flops that share the same clock and data enable inputs. Each flip-flop is positive-edge triggered by signals at the clock input (pin 7). There are four data inputs, labeled 0 to 3 (pins 11 to 14). The corresponding outputs, also labeled 0 to 3, can assume three states: low, high, and a high impedance ("Z") level that effectively "disconnects" the outputs from following circuits. Pins 1 and 2 are output enable 1 (OE1) and output enable 2 (OE2); both of these must be low or all outputs will be high impedance. Two other inputs are data enable 1 (DE1) and data enable 2 (DE2). These both must be held low to permit signals from the input to appear at the output. The reset input (pin 15) is asynchronous and must be held low; otherwise all inputs are at a low level. Maximum clock frequency is 6 MHz at 2 V and 30 MHz at 5 V. In the following logic table, "D" denotes any of the data inputs:

INPUTS							OUTPUT
OE1	OE2	Reset	Clock	DE1	DE2	D	Q
L	L	H	X	X	X	X	L
L	L	L	L	X	X	X	No change
L	L	L	H	X	X	X	No change
L	L	L	R	H	X	X	No change
L	L	L	R	X	H	X	No change
L	L	L	R	L	L	L	L
L	L	L	R	L	L	H	H
L	L	L	F	X	X	X	No change
L	H	X	X	X	X	X	High impedance
H	L	X	X	X	X	X	High impedance
H	H	X	X	X	X	X	High impedance

52 74HC174 Hex D Flip-Flop with Common Clock and Reset

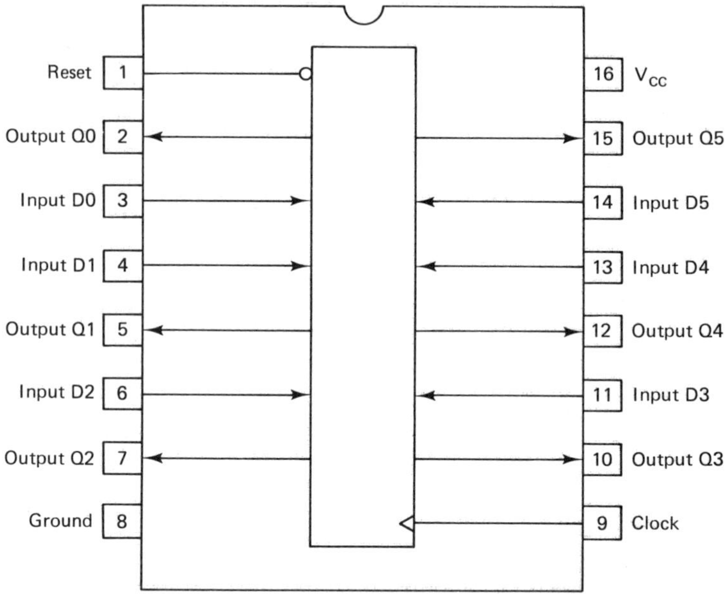

This device consists of six separate D flip-flop circuits sharing the same clock and reset inputs. The six inputs (labeled D0 to D5) have corresponding outputs (labeled Q0 to Q5); there are no complementary outputs. Each flip-flop is positive-edge triggered. There is also an asynchronous reset input that must be held high for normal operation; a low reset input sets all outputs to low. Maximum clock frequency is 6 MHz at 2 V and 30 MHz at 5 V. In the logic table for this device, "D" represents any input:

INPUTS			OUTPUT
Reset	Clock	D	
L	X	X	L
H	R	H	H
H	R	L	L
H	L	X	No change
H	F	X	No change

53 74HC175 Quad D Flip-Flop with Common Clock and Reset

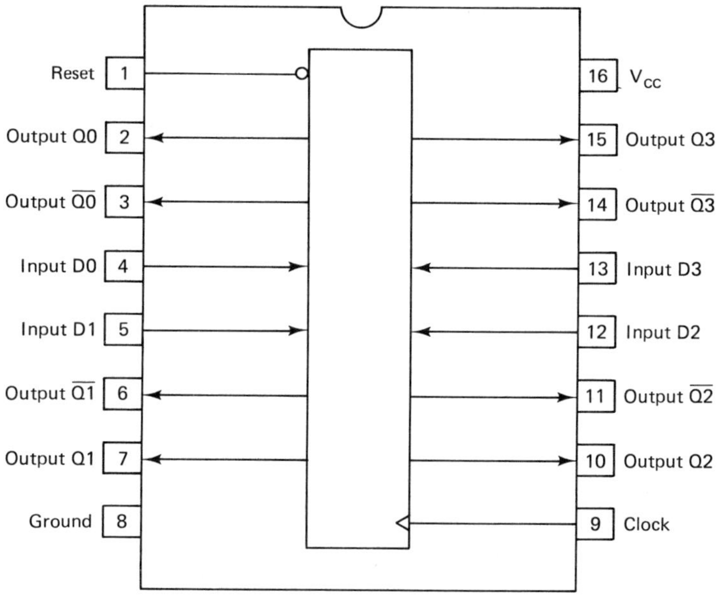

This device consists of four separate D flip-flop circuits sharing the same clock and reset inputs. The four inputs (labeled D0 to D3) have complementary outputs (labeled Q0 to Q3 and $\bar{Q}0$ to $\bar{Q}3$). There is also an asynchronous reset input that must be held high for normal operation; a low reset input sets all Q outputs to low and all $\bar{Q}$ outputs to high. Maximum clock frequency is 6 MHz at 2 V and 30 MHz at 5 V. In the logic table for this device, "D" represents any input:

INPUTS			OUTPUTS	
Reset	Clock	D	Q	$\bar{Q}$
L	X	X	L	H
H	R	H	H	L
H	R	L	L	H
H	L	X	No change	

54 74HC190 BCD Presettable Synchronous 4-Bit Up/Down Counter

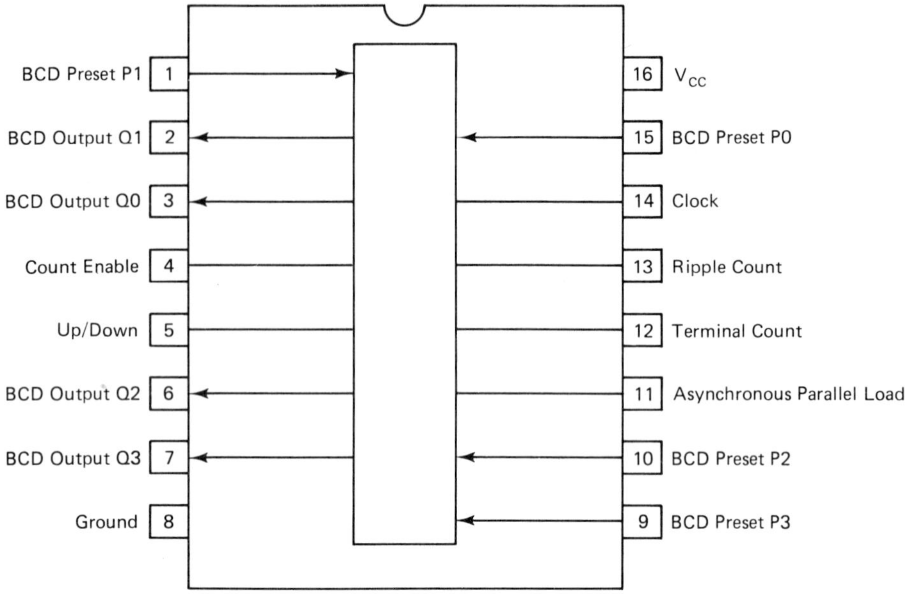

This device is a synchronous, presettable (programmable) BCD counter that can count in either direction. The direction of the counting is controlled by a signal to the up/down (U/D) input (pin 5). A low signal causes upward counting, while a high signal causes downward counting. Presetting the counter to a desired number is accomplished by a low signal to the asynchronous parallel load input (PL) at pin 11, which causes loading of the data at the BCD inputs (labeled P0 to P3). Counting action cannot take place until the count enable (CE) input at pin 4 is held low. Incrementing or decrementing of the count occurs on each positive-edge clock pulse. When an underflow or overflow of the counter occurs, the terminal count output (pin 12) goes high (it is normally low) and remains so for one clock cycle. This also causes the ripple count output (pin 13) to go low (it is normally high) and to remain so for the low portion of the clock pulse. The terminal count and ripple count output are used for cascading counters. Maximum clock frequency is 4 MHz at 2 V and 20 MHz at 5 V. The function table for this device is:

INPUTS				OPERATION
PL	CE	U/D	Clock	
H	L	L	R	Count upward
H	L	H	R	Count downward
L	X	X	X	Asynchronous preset
H	H	X	X	No change

55 74HC191 Binary Presettable 4-Bit Up/Down Counter

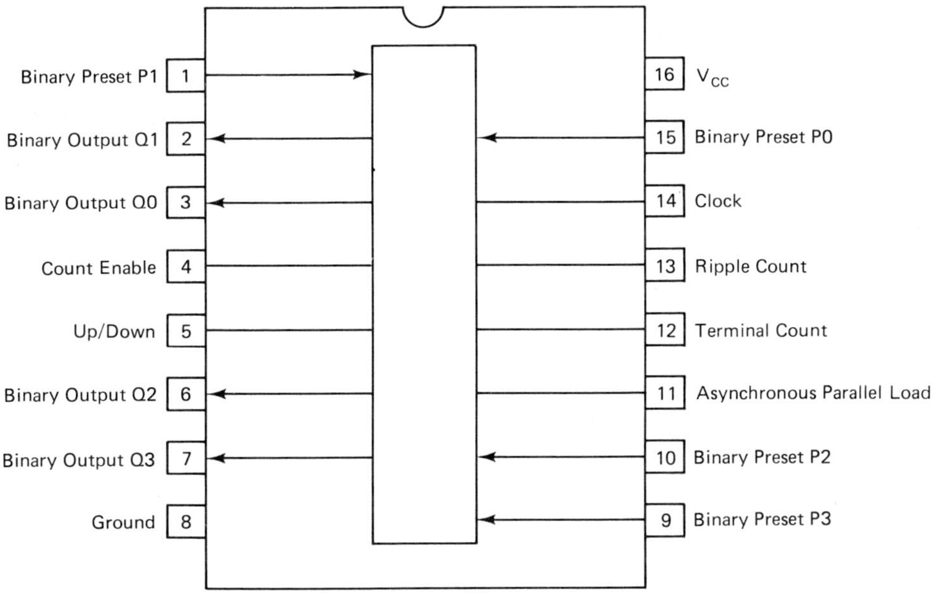

This device is a synchronous, presettable (programmable) binary counter that can count in either direction. In every respect but counting mode (binary as opposed to BCD), this device is identical to the 74HC190 BCD up/down counter. For operation, specifications, and function table, consult the listing for that device.

56 74HC194 4-Bit Bidirectional Universal Shift Register

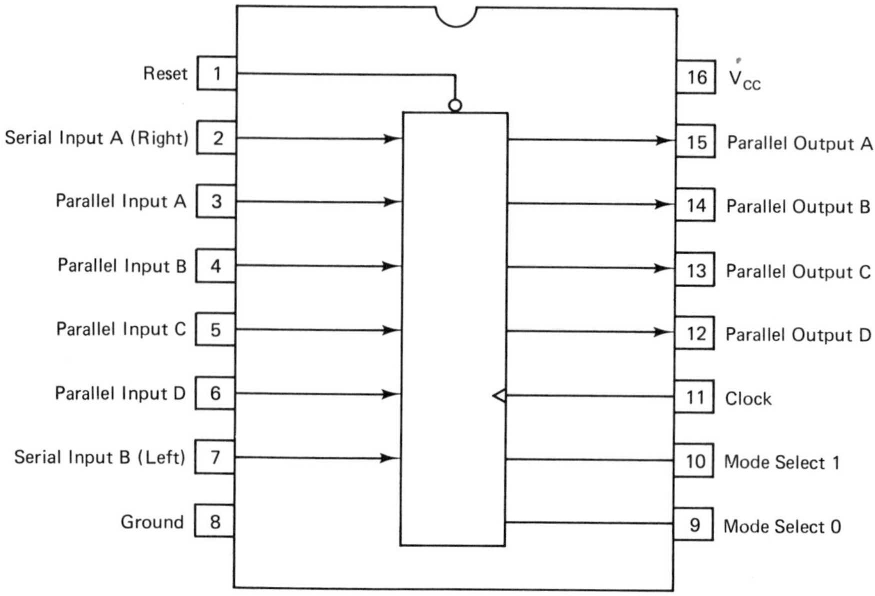

This device is a 4-bit shift register capable of shifting data to either the left or the right. There are two serial inputs, serial input A (SA) at pin 2 and serial input B (SB) at pin 7. These are also known as the right serial input and left serial input, respectively. There are also four parallel inputs (labeled A to D) and four parallel outputs (also labeled A to D). There are two mode select inputs at pin 9 (MS0) and pin 10 (MS1). In normal operation, the register shifts right when MS0 is high and MS1 is low. It shifts left when MS0 is low and MS1 is high. Data at the parallel inputs are loaded on a positive-edge clock pulse when MS0 and MS1 are both high. The data in the register are shifted one position on each positive-edge clock pulse. There is also an asynchronous reset input (pin 1), which sets all outputs to low when it is low; for normal operation, the reset input must be high. Maximum clock frequency is 6 MHz at 2 V and 30 MHz at 5 V. The function table for this device is:

INPUTS						OUTPUTS
Reset	MS0	MS1	Clock	SB	SA	
L	X	X	X	X	X	All low (reset)
H	H	H	R	X	X	Load parallel data
H	L	H	R	X	H	Shift right
H	L	H	R	X	L	Shift right
H	H	L	R	H	X	Shift left
H	H	L	R	L	X	Shift left
H	L	L	X	X	X	No change
H	X	X	L	X	X	No change
H	X	X	H	X	X	No change

57 74HC195 4-Bit Universal Shift Register

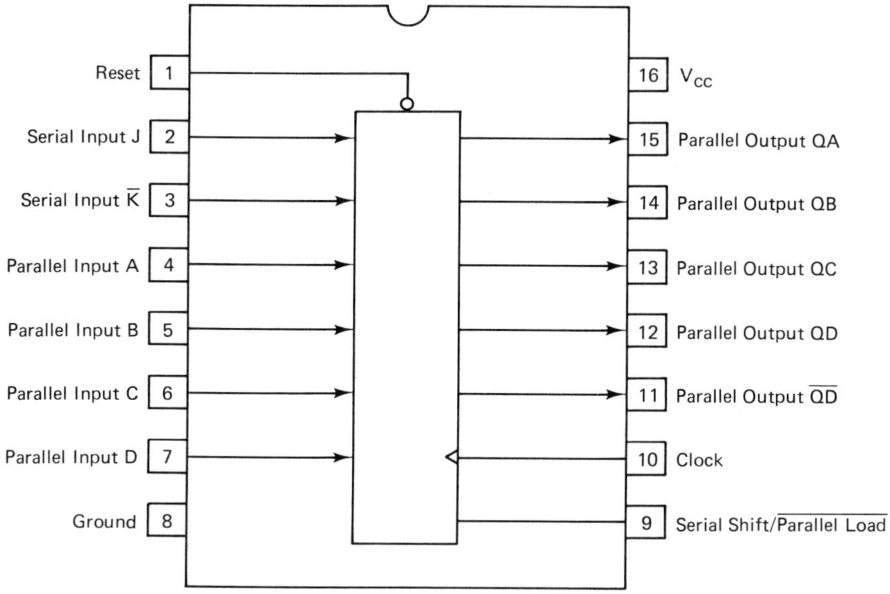

This device is a shift-right shift register with four parallel inputs (labeled A through D) and two serial inputs (labeled J and K̄). There are five parallel outputs (labeled QA through QD and its complement Q̄D̄). Loading and shifting of data are controlled by the serial shift/parallel load (S/L) input at pin 9. When this pin is low, the data at the parallel inputs are loaded into the register on the next positive-edge clock pulse. If the pin is high, the data are shifted right and appear at the five outputs. Each data shift occurs on the positive edge of each clock pulse. There is an asynchronous reset input at pin 1. For normal operation, this should be kept high; if it goes to low, all outputs go to low except for Q̄D̄, which goes to high, since it is the complement of QD. Maximum clock freequency is 6 MHz at 2 V and 30 MHz at 5 V. Maximum quiescent mode supply current is 8 μA. The function table for this device is:

INPUTS					OUTPUTS
Reset	S/L	Clock	J	K̄	
L	X	X	X	X	All low except for Q̄D̄
H	L	R	X	X	Load parallel inputs
H	H	L	X	X	Hold (no change)
H	H	R	L	H	Shift right
H	H	R	L	L	Shift right (QA low)
H	H	R	H	H	Shift right (QA high)
H	H	R	H	L	Shift right (QA toggles)

58 74HC237 1 of 8 Decoder/Demultiplexer with Address Latch

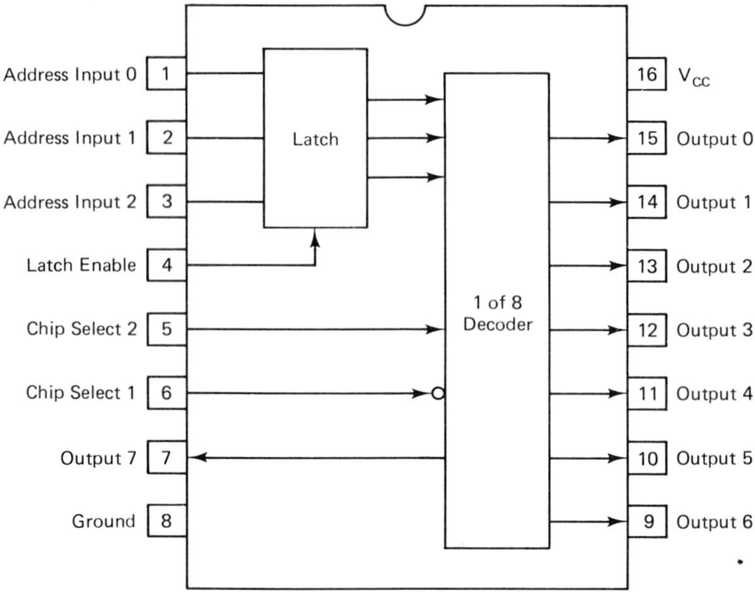

This device takes a 3-bit address code and uses it to produce a single high level from one of the eight outputs labeled 0 through 7. The address inputs are labeled 0, 1, and 2. There is an internal latch for storing the address. The internal latch is controlled by the latch enable (LE) input, which should be kept low for normal operation. There are also chip select 1 (CS1) and chip select 2 (CS2) inputs, which are provided to allow cascading of devices. For normal operation, chip select 1 should be kept high and chip select 2 should be kept low. The maximum propagation delay is 235 ns at 2 V and 47 ns at 5 V. The function table for this device is:

INPUTS						OUTPUTS							
LE	CS1	CS2	2	1	0	0	1	2	3	4	5	6	7
X	X	H	X	X	X	L	L	L	L	L	L	L	L
X	L	X	X	X	X	L	L	L	L	L	L	L	L
L	H	L	L	L	L	H	L	L	L	L	L	L	L
L	H	L	L	L	H	L	H	L	L	L	L	L	L
L	H	L	L	H	L	L	L	H	L	L	L	L	L
L	H	L	L	H	H	L	L	L	H	L	L	L	L
L	H	L	H	L	L	L	L	L	L	H	L	L	L
L	H	L	H	L	H	L	L	L	L	L	H	L	L
L	H	L	H	H	L	L	L	L	L	L	L	H	L
L	H	L	H	H	H	L	L	L	L	L	L	L	H
H	H	L	X	X	X			Holds at previous level					

59 74HC240 Octal 3-State Inverting Buffer

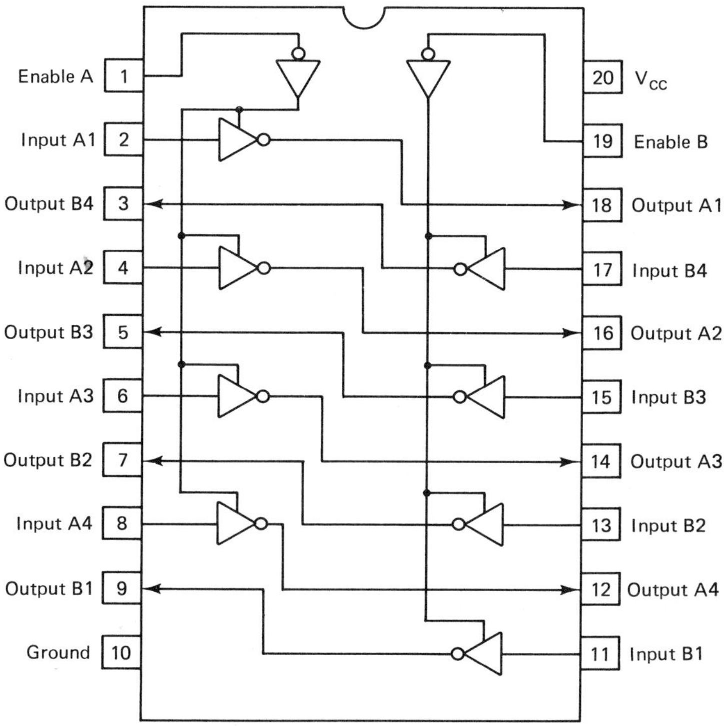

This device consists of eight separate inverting buffer gates whose operation is controlled by a signal applied to enable inputs. There are two sets of four inverting buffers; one set has inputs and corresponding outputs labeled A1 through A4, while the other has inputs and outputs labeled B1 through B4. The enable inputs are labeled A and B, and they identify which set (those labeled A or B) of inverting buffers they control. The enable signal must be low for each buffer to operate. When enabled, each buffer produces an output signal that is the opposite of its input. When the enable signal is high, the output of each buffer has a high impedance. Maximum propagation delay is 100 ns at 2 V and 20 ns at 5 V. This device is available in an HCT version. The logic table for each buffer is:

INPUTS		OUTPUT
Enable	Input	
L	L	H
L	H	L
H	X	Z

60 74HC241 Octal 3-State Noninverting Buffer

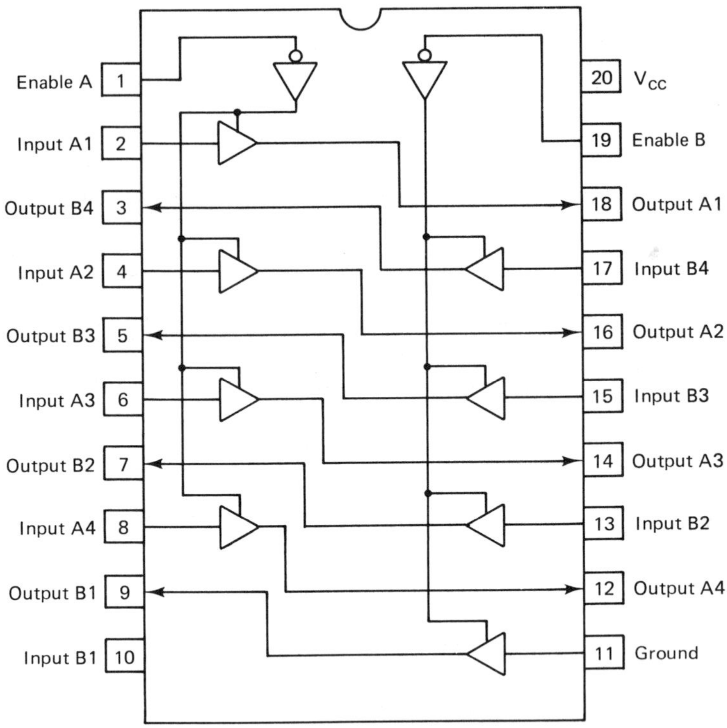

This device consists of eight separate noninverting buffers whose operation is controlled by enable signals. This device is similar to the 74HC240, but there are some major differences. All of these buffers are noninverting, meaning that the output of each will be the same logic level as the input. While buffers are grouped into two sets labeled A and B, the enable signals operate differently for them. In the A group of four buffers, a low signal will enable the gates to operate normally, while a high signal will give the outputs a high impedance. By contrast, a high logic level signal will enable the B group of buffers, while a low signal will produce a high impedance output. This device is available in HCT versions. Maximum propagation delay is 115 ns at 2 V and 23 ns at 5 V. The logic table for each group of buffers is:

"A" INPUTS		"A" OUTPUT
Enable—A	Input—A	
L	L	L
L	H	H
H	X	Z

"B" INPUTS		"B" OUTPUT
Enable—B	Input—B	
H	L	L
H	H	H
L	X	Z

61 74HC242 Quad 3-State Bus Transceiver with Inverting Outputs

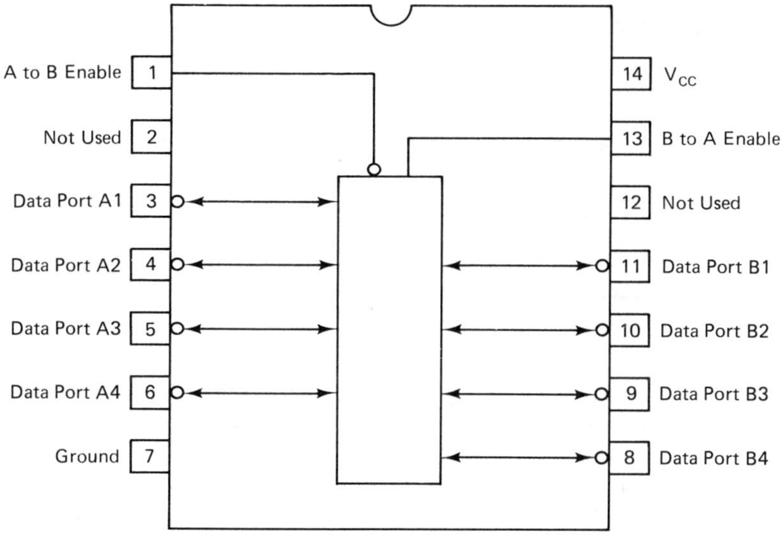

This device allows communications between two different four-bit data buses (labeled A and B). Communications may be from A to B or from B to A depending on the signals applied to two enable pins. The A to B (A/BE) enable pin is at pin 1, while the B to A (B/AE) enable pin is at pin 13. Both enable inputs must have the same logic-level input for normal operation; if the inputs differ, the result is that all inputs and outputs have a high impedance and the buses are isolated from each other. This device is an inverting transceiver, meaning that the signals from the "input bus" appear at the outputs inverted from their levels at input. Maximum propagation delay is 100 ns at 2 V and 20 ns at 5 V. Maximum quiescent mode current consumption is 8 μA. In the function table for this device, "I" represents the input bus while "O" represents the inverted output:

A/BE	B/AE	PORT—A	PORT—B
H	H	O	I
L	H	Z	Z
H	L	Z	Z
L	L	I	O

62 74HC243 Quad 3-State Bus Transceiver with Noninverting Outputs

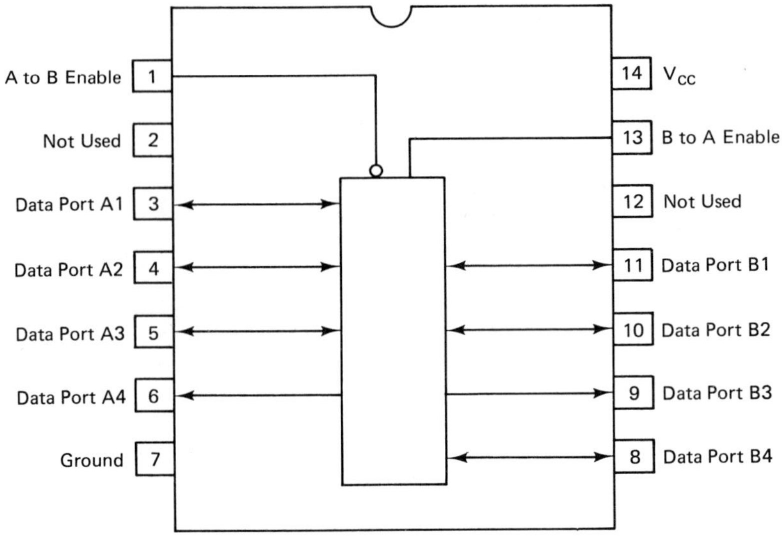

This device is very similar to the 74HC242, but the outputs are noninverting; signals are transferred between the two buses without any changes in logic level. Specifications and operation of the 74HC243 are otherwise the same as for the 74HC242. In the function table for this device, "I" represents the input bus while "O" represents the signals output to the other bus:

A/BE	B/AE	PORT—A	PORT—B
H	H	O	I
L	H	Z	Z
H	L	Z	Z
L	L	I	O

63 74HC244 Octal 3-State Noninverting Buffer

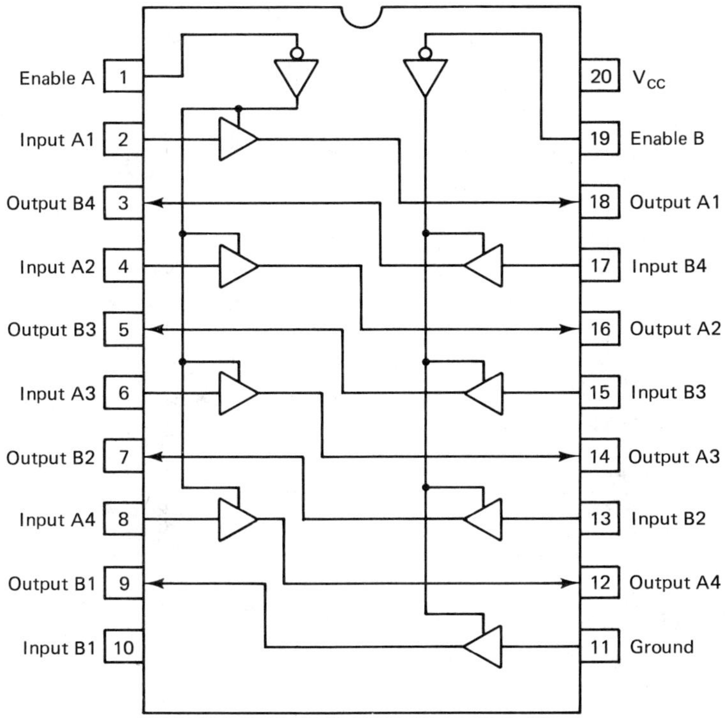

This device contains two "banks" of four noninverting buffers. The two sets of buffers are labeled A and B, with numerical suffixes identifying the individual inputs and outputs. Two enable inputs, also labeled A and B, control the operation of the buffers in each respective bank. When the enable input of a given bank is low, the output of any buffer in that bank will follow its input. If the enable is high, the output will be high impedance regardless of the input. This device is available in an HCT version. Propagation delay is 115 ns at 2 V and 23 ns at 5 V. Maximum quiescent mode power consumption is 8 μA per gate. The function table for each gate is:

ENABLE	INPUT	OUTPUT
L	L	L
L	H	H
H	X	Z

64 74HC245 Octal 3-State Noninverting Bus Transceiver

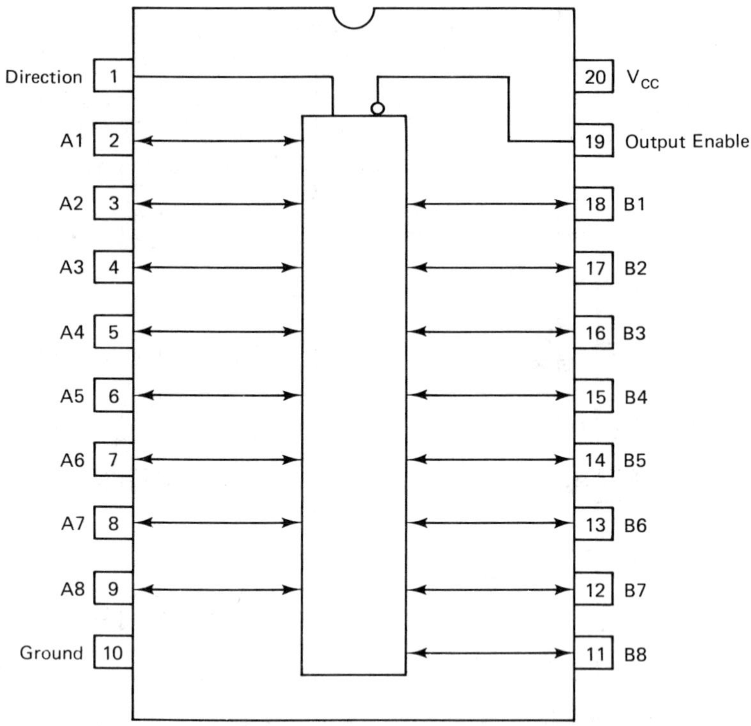

This device is used for asynchronous two-way communication between eight-bit bus lines. The direction of the data transfer depends on the signal applied to the direction input (pin 1). A low signal to this input allows data to flow from bus B to bus A, while a high signal causes data to travel from bus A to bus B. There is also an output enable input (pin 19). This input must be held at a low level for proper operation; if it goes high, the buses are isolated from each other. The maximum output transition time (i.e., time required to change direction of data flow) is 60 ns at 2 V and 12 ns at 5 V. Quiescent mode power consumption is 8 μA. This device is also available in an HCT version. The function table for this device is:

ENABLE	DIRECTION	OPERATION
L	L	Data move from bus B to bus A
L	H	Data move from bus A to bus B
H	X	Buses isolated from each other

65 74HC251 8-Input Data Selector/Multiplexer with 3-State Outputs

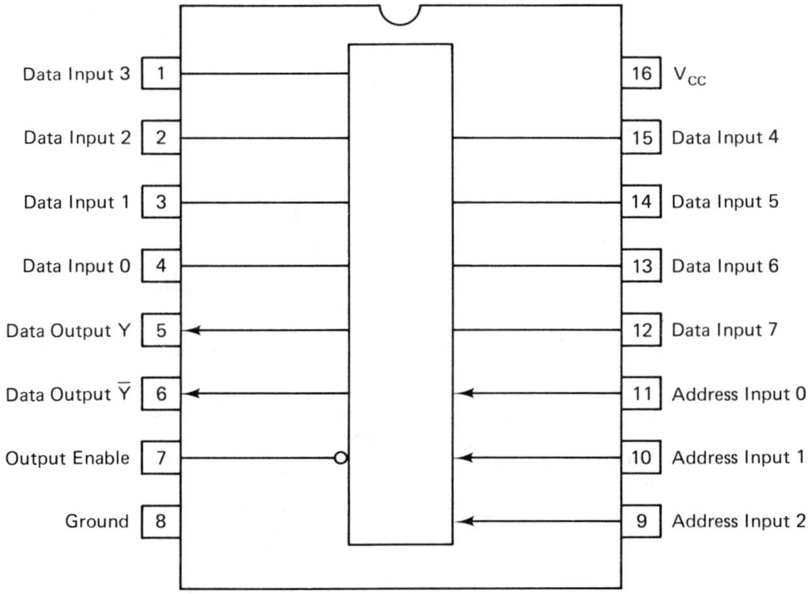

This device selects one of eight different binary inputs, as determined by the three address inputs, and the selected input appears at output Y (pin 5) and its complement at output $\overline{Y}$ (pin 6). There is an output enable input (pin 7), which must be kept at a low level for normal operation. If the output enable is high, both Y and $\overline{Y}$ will be at a high impedance. Maximum propagation delay is 185 ns at 2 V and 37 ns at 5 V. In the following function table, the numbers under the Y and $\overline{Y}$ output headings indicate the logic level present at the corresponding data input. For example, "0" represents data input 0 (pin 4). In the same manner, the letter "A" followed by a numeral indicates an address input (pins 9, 10, and 11).

INPUTS				OUTPUTS	
A2	A1	A0	ENABLE	Y	$\overline{Y}$
X	X	X	X	Z	Z
L	L	L	L	0	0
L	L	H	L	1	1
L	H	L	L	2	2
L	H	H	L	3	3
H	L	L	L	4	4
H	L	H	L	5	5
H	H	L	L	6	6
H	H	H	L	7	7

66 74HC253 Dual 4-Input Data Selector/Multiplexer with 3-State Outputs

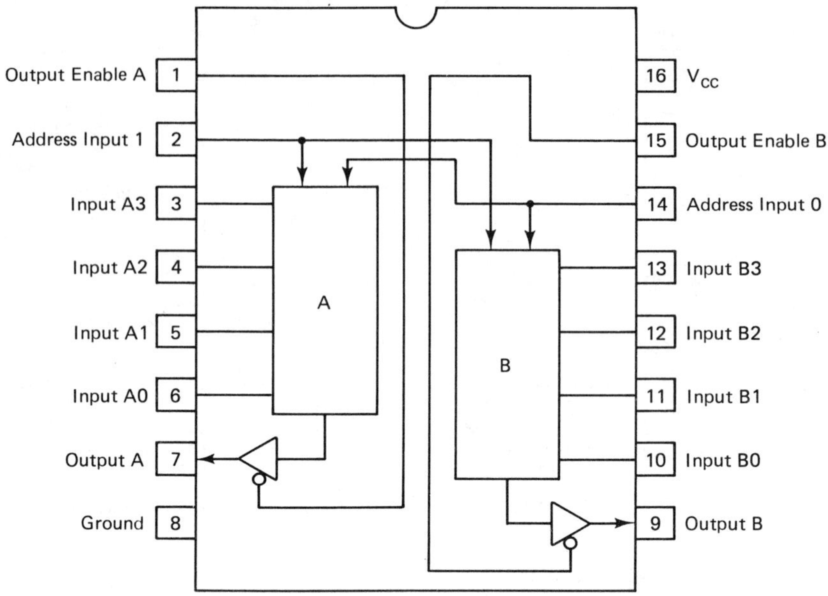

This device consists of two circuits, labeled A and B, with four separate data inputs and output enable inputs but which share two address inputs. Depending on the signals present at the address inputs (identified as A1 or A0 in the function table), one of the four data inputs is selected and appears at output A or B (pin 7 or 9). For normal operation, the corresponding output enable input (pin 1 or 15) must be kept low; if it is high, the corresponding output will have a high impedance. The maximum propagation delay is 140 ns at 2 V and 28 ns at 5 V. Quiescent mode power consumption is 8 μA per package. The following function table is for each circuit, with the numeral under the output heading representing the data input selected (i.e., 0 represents either input A0 or input B0) to appear at the corresponding output.

INPUTS			OUTPUT
A1	A0	ENABLE	
X	X	H	Z
L	L	L	0
L	H	L	1
H	L	L	2
H	H	L	3

67 74HC257 Quad 2-Input Data Selector/Multiplexer with 3-State Outputs

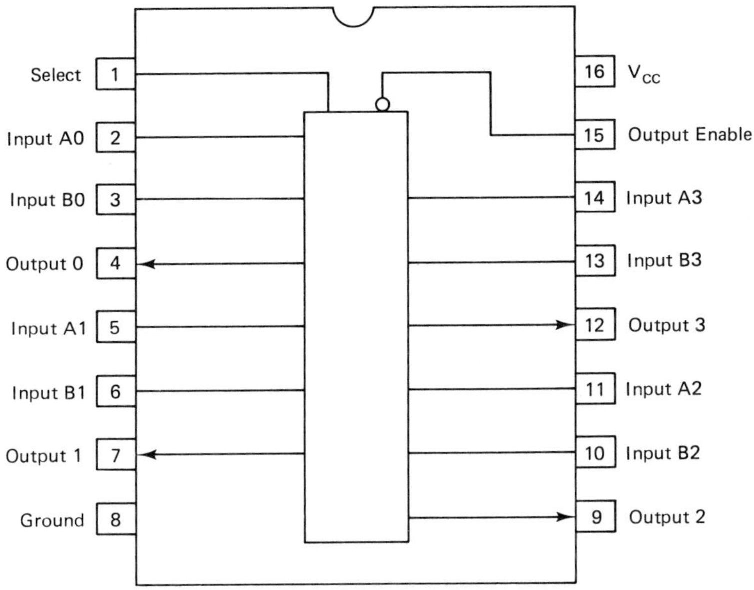

This device selects one of two four-bit inputs (labeled A and B, with numerical suffixes from 0 to 3) and allows the selected input to appear at the outputs labeled 0 through 3. The select input (pin 1) chooses the inputs sent to the outputs; a low input signal selects the A inputs, while a high input signal selects the B inputs. There is also an output enable input (pin 15), which should be kept low for normal operation. If it is high, the outputs will all have a high impedance. The outputs will correspond to the selected inputs. For example, if the A input is selected, the logic level at input A0 (pin 2) will appear at output 0 (pin 4). Propagation delay is 100 ns at 2 V and 20 ns at 5 V. The function table for this device is:

ENABLE	SELECT	OUTPUTS
H	X	All high impedance (Z)
L	L	A inputs
L	H	B inputs

68 74HC259 8-Bit Addressable Latch/1 of 8 Decoder

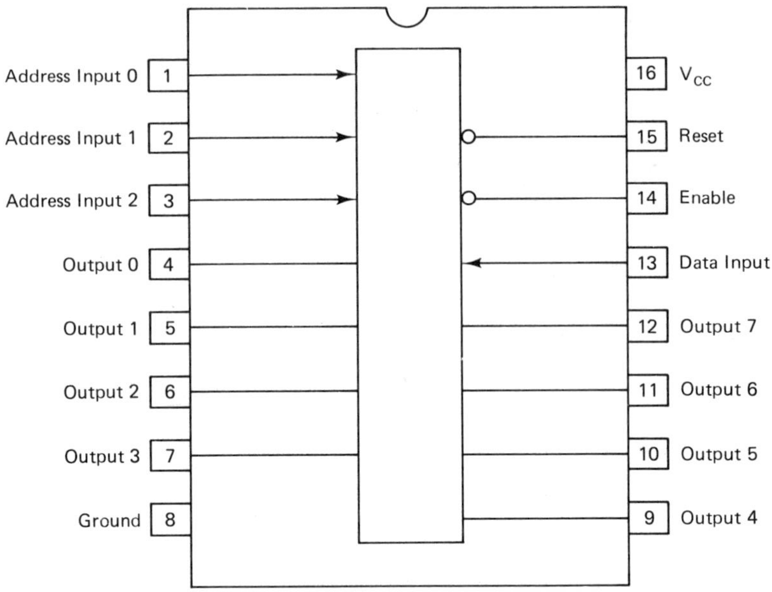

This device has four modes of operation, selected by signals applied to the enable (pin 14) and reset (pin 15) inputs. The first mode of operation is as an addressable latch, in which the signal at the data input (pin 13) is sent to the output addressed; this output follows changes in the input signal while all other outputs remain at their previous state. The second mode of operation is as a memory, in which all outputs retain their previous states and are unaffected by the data or address inputs. In the decoder (demultiplexer) mode, the addressed output follows the data input signal while all other inputs remain low. The fourth mode is the reset mode, in which all outputs are low and are unaffected by the address and data inputs. The following table shows how the enable and reset inputs select operating mode:

ENABLE	RESET	MODE
L	H	Addressable latch
H	H	Memory
L	L	Decoder
H	L	Reset

The output selected depends on signals applied to the three address inputs labeled 0, 1, and 2. The maximum propagation delay from the data input to any of the outputs is 185 ns at 2 V and 37 ns at 5 V. The maximum quiescent mode current consumption is 8 μA. The following table shows the address input signals that select a desired output:

ADDRESS INPUTS 2	1	0	OUTPUT
L	L	L	0
L	L	H	1
L	H	L	2
L	H	H	3
H	L	L	4
H	L	H	5
H	H	L	6
H	H	H	7

69 74HC273 Octal D Flip-Flop with Common Clock and Reset

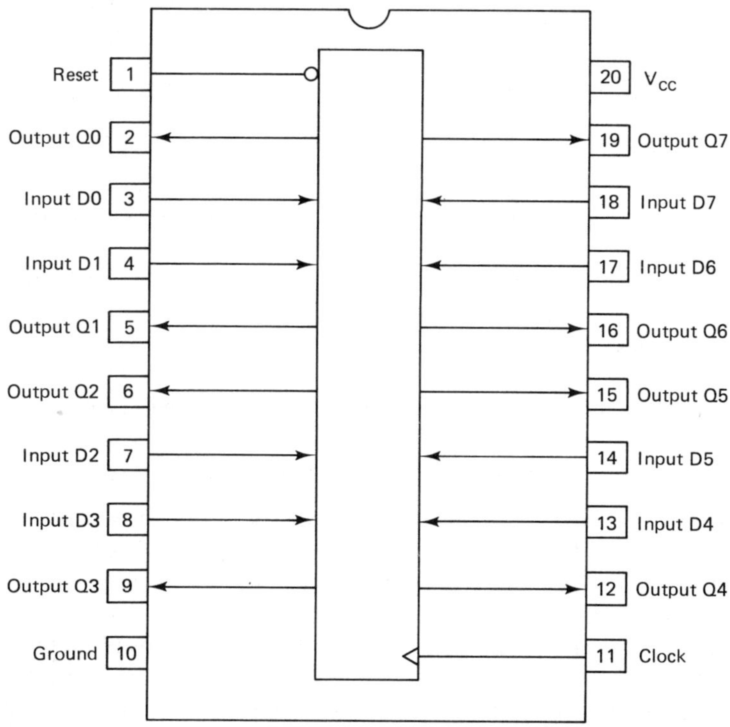

This device contains eight separate D flip-flops that share common clock (pin 11) and reset (pin 1) inputs. The reset is asynchronous and sets all outputs to low when it is low; for normal operation, the reset input must be kept high.

Each flip-flop is triggered on the rising edge of a clock pulse. The maximum clock frequency is 6 MHz at 2 V and 30 MHz at 5 V. The logic table for each device is:

INPUTS			OUTPUT
Reset	Clock	D	
L	X	X	L
H	R	H	H
H	R	L	L
H	L	X	No change
H	F	X	No change

70 74HC280 9-Bit Odd/Even Parity Generator/Checker

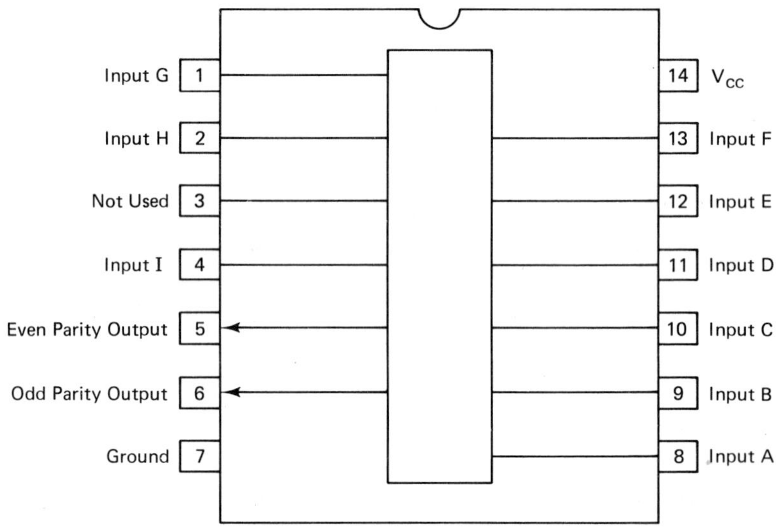

This device accepts a nine-bit data word and indicates whether it has odd or even parity. The nine inputs are labeled A through I with an even parity output (pin 5) and an odd parity output (pin 6). If the number of inputs that are high is an even number (0, 2, 4, 6, or 8), the even parity output will be high and the odd parity output will be low. If the number of inputs that are high is an odd number (1, 3, 5, 7, or 9), the even parity output will be low and the odd parity output will be high. Words greater than nine bits can be parity-checked by cascading several 74HC280 devices.

71 74HC283 4-Bit Binary Full Adder with Fast Carry

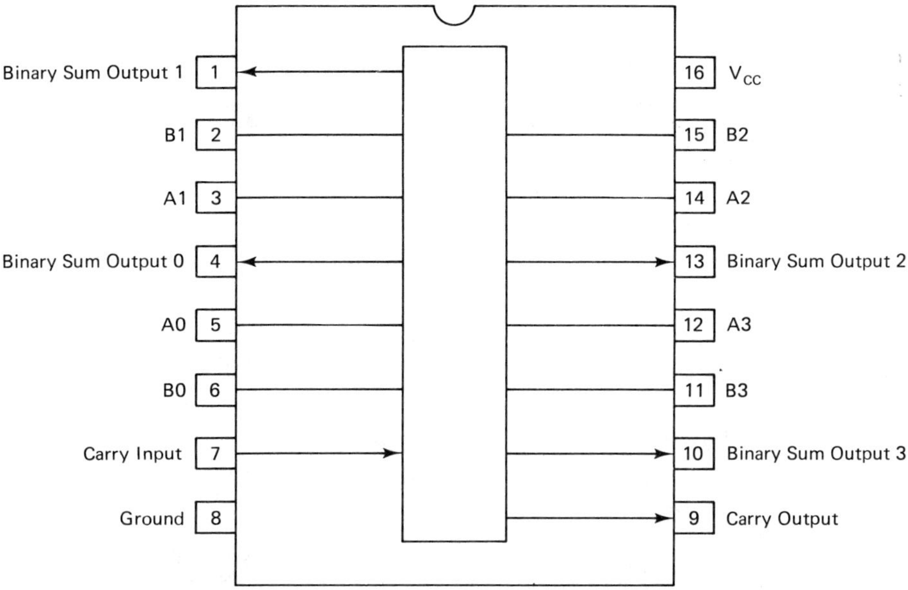

This device adds two four-bit words, plus any bit present at the carry input (pin 7), and outputs the resulting binary sum at the outputs labeled 0 through 3. There is also a carry output at pin 9 for "overflow" from the binary sum outputs. The most significant bit of each input and output is the one labeled 3 (A3, B3, and binary sum output 3). This device can be used either with all inputs and outputs active on high logic levels (positive logic) or with all inputs and outputs active on low logic levels (negative logic). If positive logic is used, the carry input must be held low when no carry input is intended. Several of these devices may be cascaded using the carry input and carry output of each device. Propagation delay is a maximum of 210 ns at 2 V and 42 ns at 5 V. Maximum quiescent mode current is 8 μA.

72 74HC299 8-Bit Bidirectional Universal Shift Register with Parallel Input/Output

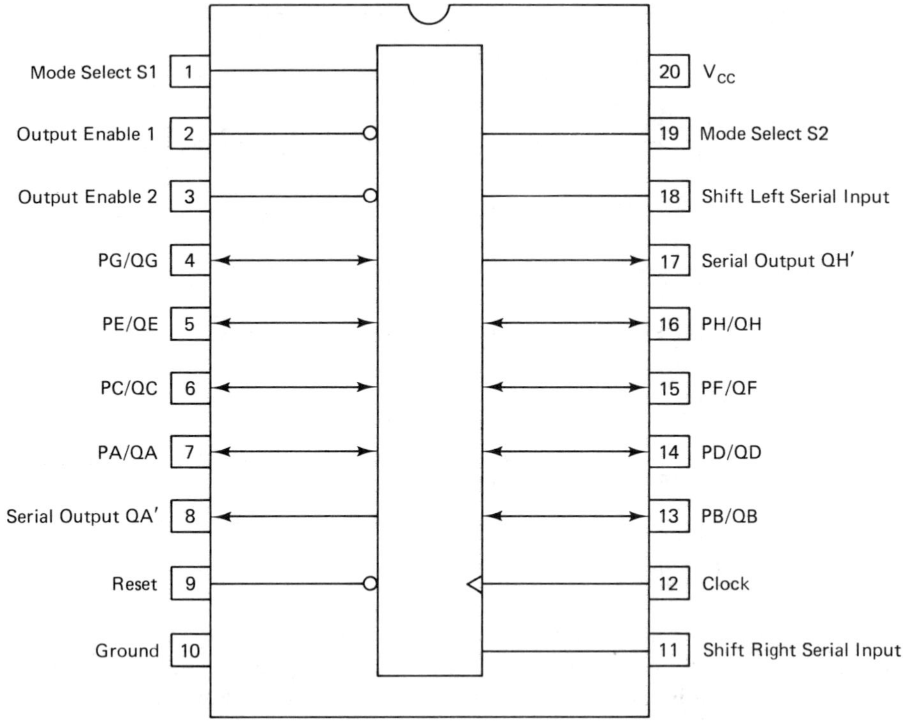

This device is a shift register capable of handling eight bits and shifting them to the left or right on the positive edge of a clock input pulse. There is an eight-bit parallel data port, labeled PA/QA through PH/QH, which can be used for input or output. There are two serial inputs, labeled shift right (SR) and shift left (SL), and two serial data outputs. There are two mode select inputs (S1 and S2) and two output enable inputs (OE1 and OE2), which select the desired mode of operation. If both mode select inputs are high, the parallel data port serves as an input. There is also an asynchronous reset input at pin 9, which sets all outputs to low when it is low. Maximum clock frequency is 5 MHz at 2 V and 25 MHz at 5 V. The function table for this device is:

INPUTS								OPERATION
Reset	S2	S1	OE1	OE2	Clock	SR	SL	
L	X	L	L	L	X	X	X	All outputs low
L	L	X	L	L	X	X	X	All outputs low
L	H	H	X	X	X	X	X	All outputs high impedance

INPUTS								OPERATION
Reset	S2	S1	OE1	OE2	Clock	SR	SL	
H	L	H	H	H	R	D	X	Shift right
H	L	H	X	H	R	D	X	Shift right
H	L	H	L	L	R	D	X	Shift right
H	H	L	H	X	R	X	D	Shift left
H	H	L	X	H	R	X	D	Shift left
H	H	L	L	L	R	X	D	Shift left
H	H	H	X	X	R	X	X	Parallel load
H	L	L	H	X	X	X	X	Hold
H	L	L	X	H	X	X	X	Hold
H	L	L	L	L	X	X	X	Hold

73 74HC354 8-Input Data Selector/Multiplexer with Data and Address Latches and 3-State Outputs

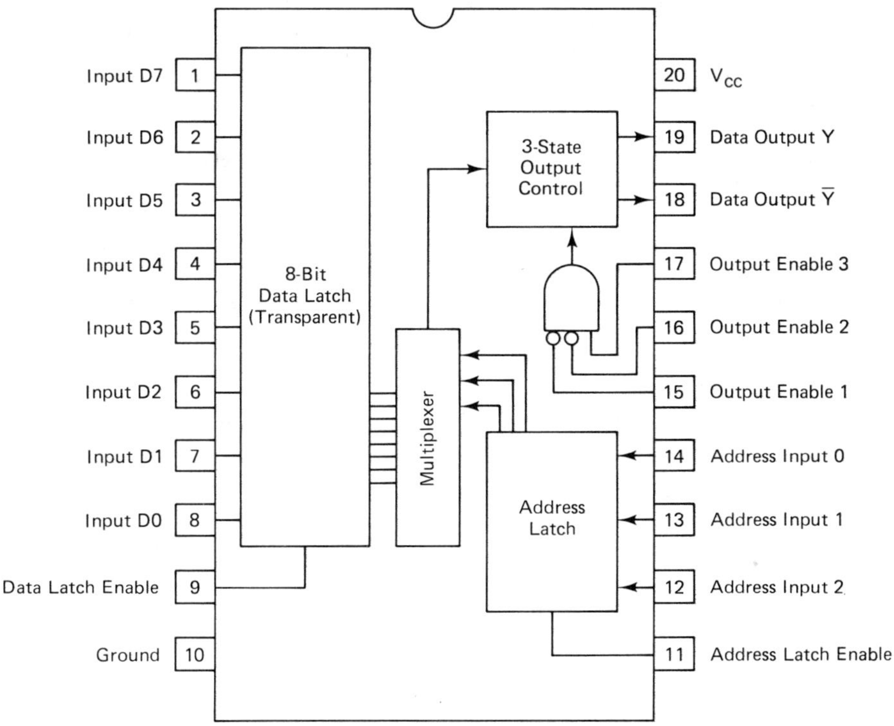

This device has eight latched data inputs, labeled D0 through D7, and selects one of the inputs according to the three address inputs. The selected input appears at output Y (pin 19) and its complement at output $\overline{Y}$ (pin 18).

The input signals are stored in an internal eight-bit transparent latch when the data latch enable (DLE) input at pin 9 is high. When pin 9 is high, any changes in the input signals are ignored. If pin 9 is low, the latch is transparent and changes in the inputs are reflected in the output. The address inputs may likewise be stored in an internal address latch when the address latch enable (pin 11) is high. If pin 11 is high, changes in the address inputs are ignored, while a low pin 11 will cause the device to respond immediately to address changes. There are three output enable inputs (labeled OE1, OE2, and OE3), which can be used to set both outputs to a high impedance. Maximum propagation delay is 210 ns at 2 V and 42 ns at 5 V. In the following function table, A2, A1, and A0 represent the corresponding address inputs:

INPUTS							OUTPUTS	
A2	A1	A0	DLE	OE1	OE2	OE3	Y	$\bar{Y}$
X	X	X	X	H	X	X	Z	Z
X	X	X	X	X	H	X	Z	Z
X	X	X	X	X	X	L	Z	Z
L	L	L	L	L	L	H	D0	D0
L	L	H	L	L	L	H	D1	D1
L	H	L	L	L	L	H	D2	D2
L	H	H	L	L	L	H	D3	D3
H	L	L	L	L	L	H	D4	D4
H	L	H	L	L	L	H	D5	D5
H	H	L	L	L	L	H	D6	D6
H	H	H	L	L	L	H	D7	D7

74 74HC356 Clocked 8-Input Data Selector/Multiplexer with Data and Address Latches and 3-State Outputs

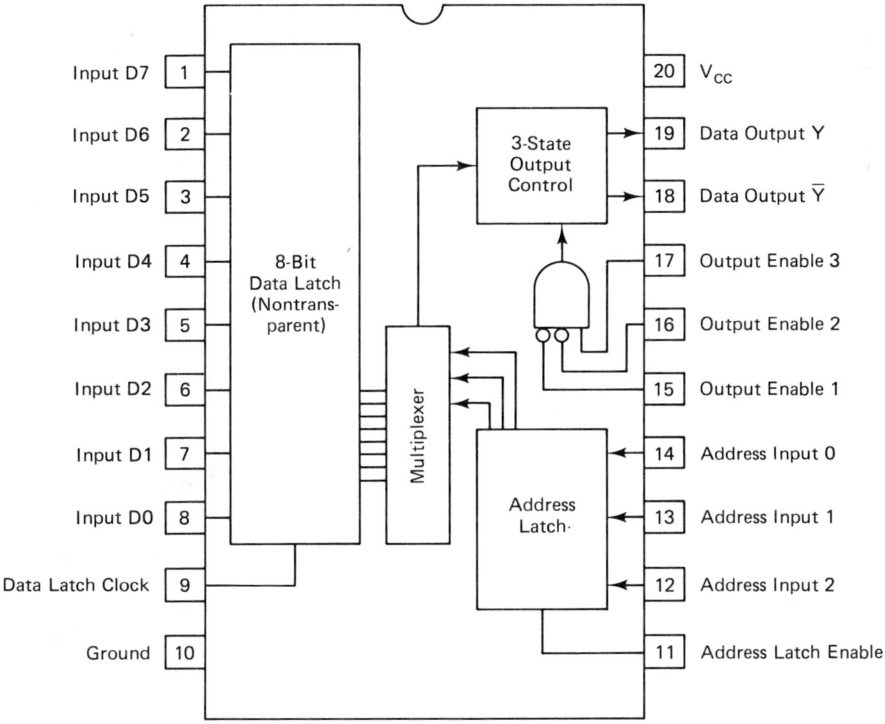

This device is similar to the 74HC354, selecting one of the eight input signals and allowing it and its complement to appear at outputs Y and $\overline{Y}$. However, the inputs are latched into the internal nontransparent latch on the positive edge of a clock input signal applied to pin 9. If the clock input is a negative edge of the level transition, or is a constant high or low level, no changes in the inputs are loaded into the latch and the contents of the latch are unaffected. The other inputs function in the same manner as the 74HC354. The function table for this device is as follows:

INPUTS							OUTPUTS	
A2	A1	A0	Clock	OE1	OE2	OE3	Y	$\overline{Y}$
X	X	X	X	H	X	X	Z	Z
X	X	X	X	X	H	X	Z	Z
X	X	X	X	X	X	L	Z	Z
L	L	L	R	L	L	H	D0	D0
L	L	H	R	L	L	H	D1	D1

INPUTS							OUTPUTS	
A2	A1	A0	Clock	OE1	OE2	OE3	Y	$\bar{Y}$
L	H	L	R	L	L	H	D2	D2
L	H	H	R	L	L	H	D3	D3
H	L	L	R	L	L	H	D4	D4
H	L	H	R	L	L	H	D5	D5
H	H	L	R	L	L	H	D6	D6
H	H	H	R	L	L	H	D7	D7
X	X	X	F	X	X	X	No change	
X	X	X	H	X	X	X	No change	
X	X	X	L	X	X	X	No change	

75 74HC365 Hex 3-State Noninverting Buffer with Common Enables

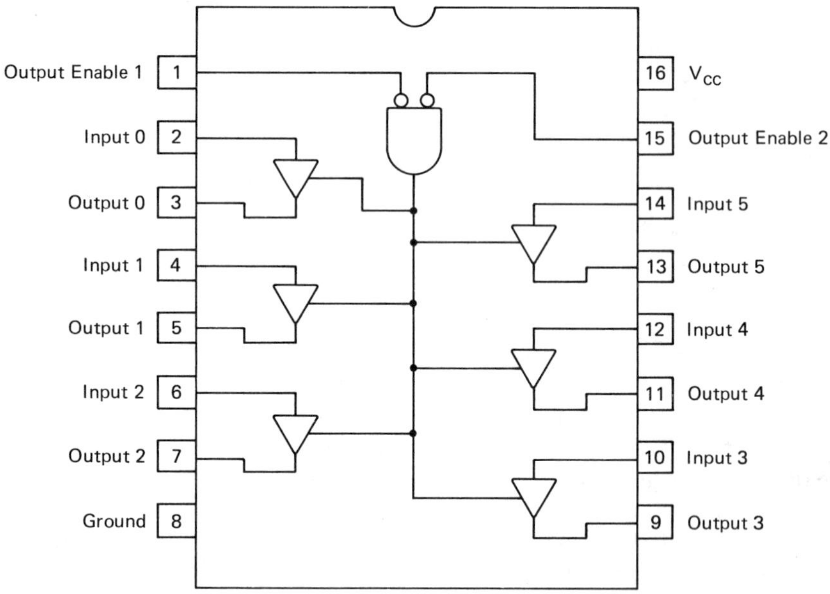

This device consists of six independent noninverting buffers that share two common output enable inputs (labeled OE1 and OE2 in the logic table below). If both OE1 and OE2 are low, any signals present at the input of a buffer appear at the output of that buffer. When either or both of the output enable inputs are high, the outputs of all buffers will be at a high impedance. The maximum propagation delay is 120 ns at 2 V and 24 ns at 5 V. Input capacitance is 10 pF. The logic table for each buffer is:

INPUTS			OUTPUT
OE1	OE2	Input	
L	L	L	L
L	L	H	H
H	X	X	Z
X	H	X	Z

76 74HC366 Hex 3-State Inverting Buffer with Common Enables

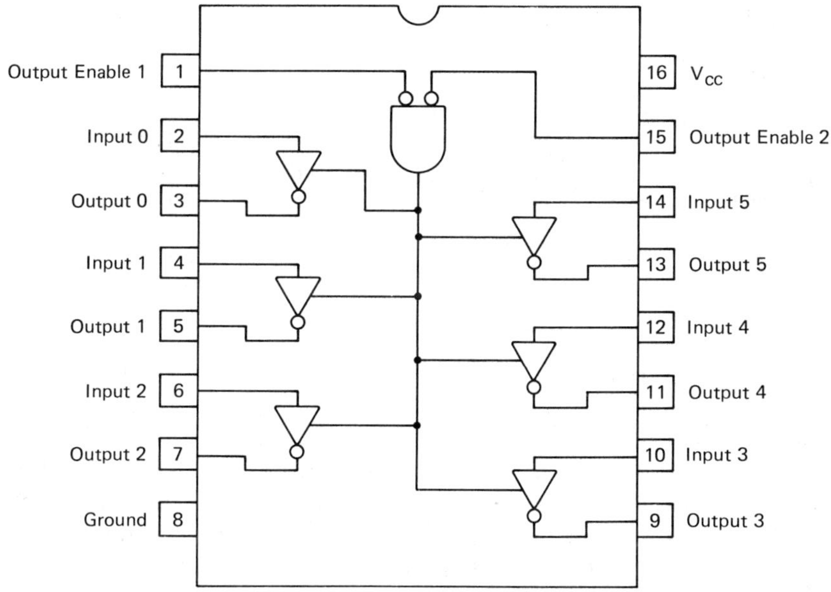

This device is very similar to the 74HC365, but the input to each buffer is inverted at the output. Maximum propagation delay is 95 ns at 2 V and 19 ns at 5 V. The logic table for each buffer is:

INPUTS			OUTPUT
OE1	OE2	Input	
L	L	L	H
L	L	H	L
H	X	X	Z
X	H	X	Z

77 74HC367 Hex 3-State Noninverting Buffer with Separate 2-Bit and 4-Bit Sections

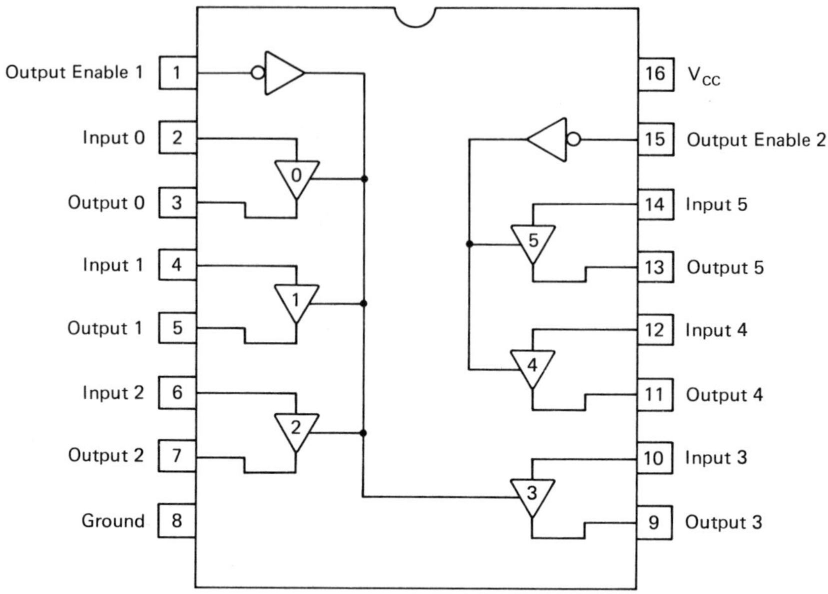

This device consists of six noninverting buffers arranged into two-bit and four-bit sections. Each section has its own output enable (OE) input. Output enable 1 controls buffers 0 through 3, while output enable 2 controls buffers 4 and 5. When an output enable input is held at a high level, the affected group of buffers all have high impedance outputs. If the output enable is at a low level, the output of the affected buffers will have the same logic levels as the input. Maximum propagation delay is 120 ns at 2 V and 24 ns at 5 V. Maximum quiescent mode current consumption is 8 μA per package. The logic table for this device is:

INPUTS		OUTPUT
OE	Input	
L	L	L
L	H	H
H	X	Z

78 74HC368 Hex 3-State Inverting Buffer with Separate 2-Bit and 4-Bit Sections

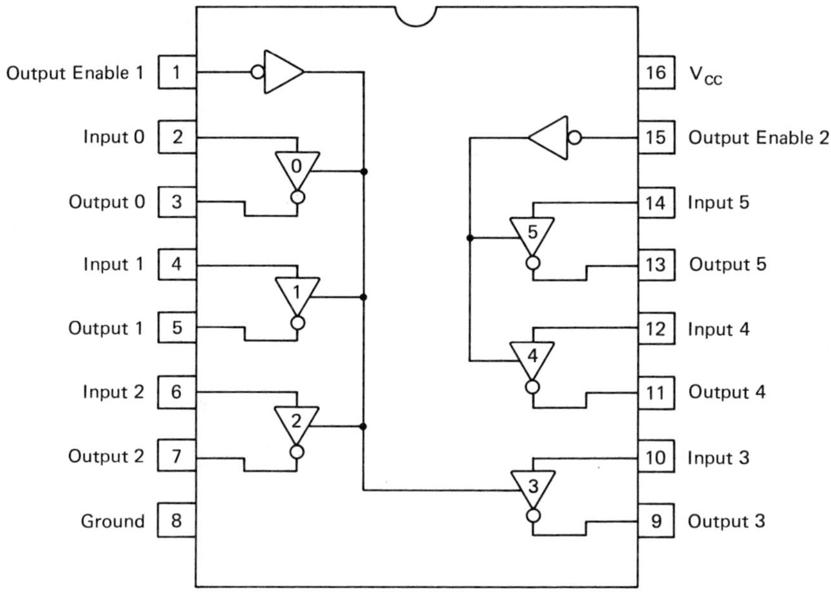

This device is very similar to the 74HC367 except that the input to each buffer is inverted at the output. Maximum propagation delay is 95 ns at 2 V and 19 ns at 5 V. The logic table for this device is:

INPUTS		OUTPUT
OE	Input	
L	L	H
L	H	L
H	X	Z

79 74HC373 Octal 3-State Noninverting Transparent Latch

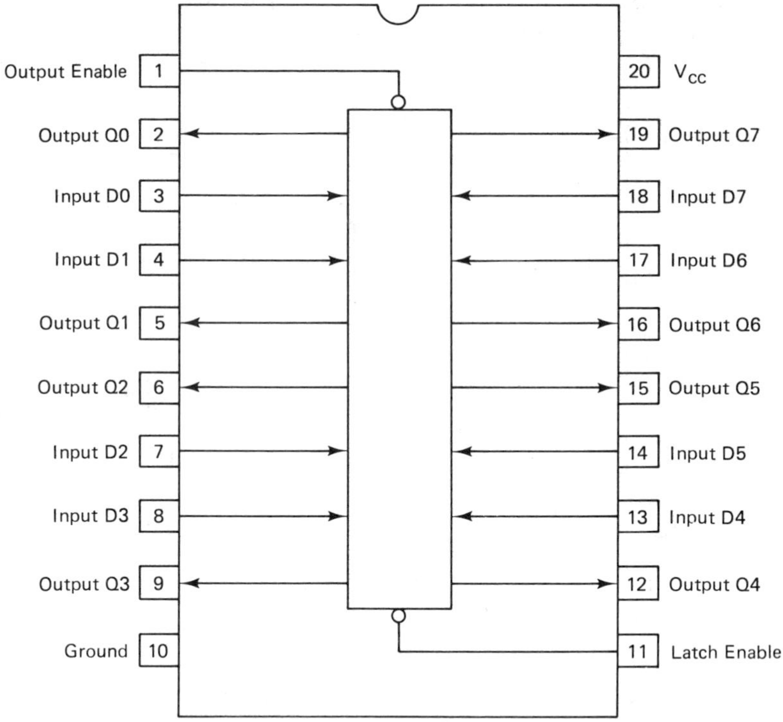

This device has eight data inputs (labeled D0 through D7) and eight corresponding data outputs (labeled Q0 through Q7). When the latch enable (LE) input is high, the outputs will follow any changes in the logic levels at the inputs. When the latch enable is low, however, the output logic levels become latched and do not change regardless of any changes in the input logic levels. The output enable (OE) input should be kept low for normal operation. When it is made high, the outputs all have a high impedance. The data stored in the latch are not affected by making the output enable high; the latched data can be made to appear at the output merely by returning the output enable logic level to low. Maximum propagation delay is 150 ns at 2 V and 30 ns at 5 V. This device is available in an HCT version. The function table for this device is:

INPUTS			OUTPUT
OE	LE	Input	
L	H	H	H
L	H	L	L
L	L	X	No change
H	X	X	Z

80 74HC374 Octal 3-State Noninverting D Flip-Flop

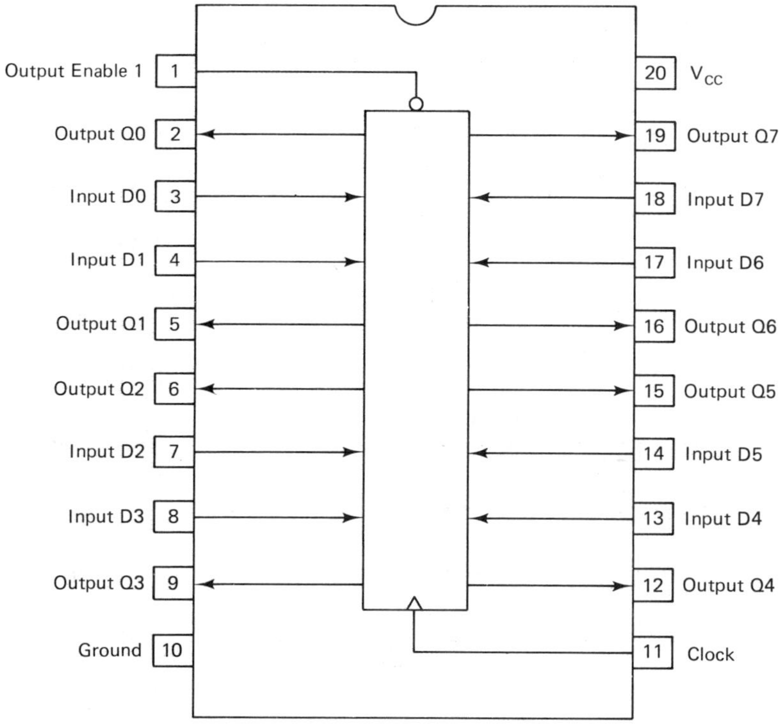

This device consists of eight separate D flip-flops with eight data (D) inputs labeled D0 through D7 and eight noninverting outputs labeled Q0 through Q7. Signals at the D inputs are transferred to the corresponding outputs on the positive edge of each clock pulse. There is an output enable (OE) input at pin 1. The output enable does not affect the states of the flip-flops; instead it sets all outputs to a high impedance level when it is at a high logic level. For normal operation, the output enable input should be kept low. The maximum clock frequency is 6 MHz is 2 V and 30 MHz at 5 V. Maximum propagation delay is 180 ns at 2 V and 36 ns at 5 V. This device is available in an HCT version. The function table for each flip-flop is:

INPUTS			OUTPUT
OE	Clock	D	
L	R	H	H
L	R	L	L
L	L	X	No change
L	H	X	No change
L	F	X	No change
H	X	X	Z

81 74HC377 Octal Noninverting D Flip-Flop with Output Enable

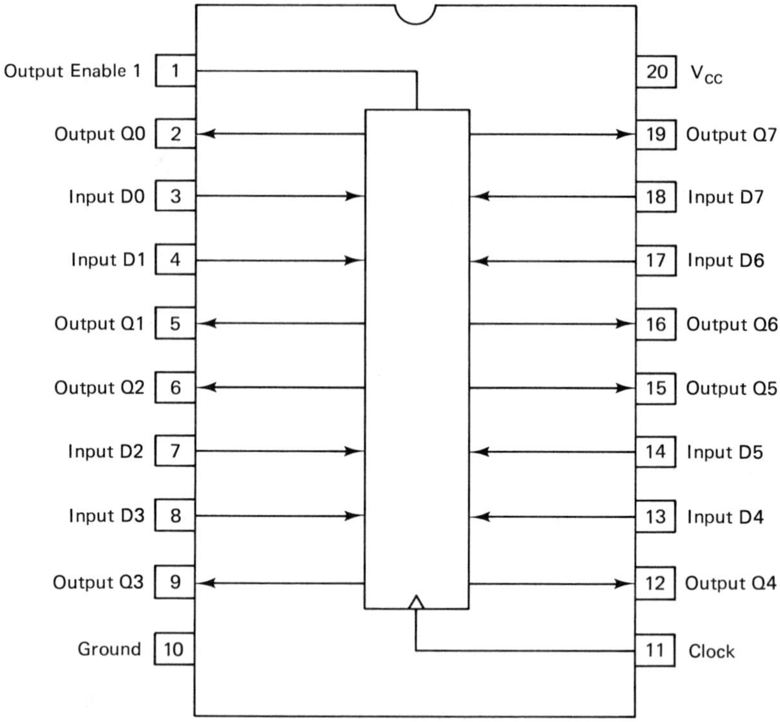

This device is similar to the 74HC374 except that a high impedance output mode is not provided. There are eight internal D flip-flops, with eight inputs labeled D0 through D7 and eight corresponding outputs labeled Q0 through Q7. Signals at the inputs are transferred to the outputs on each positive edge of a clock pulse. For normal operation, the output enable (OE) input at pin 1 must be kept low. If it is high, there will be no change in the output logic levels regardless of any changes in the inputs. Maximum clock frequency is 6 MHz at 2 V and 30 MHz at 5 V. This device is available in an HCT version. The function table for this device is:

INPUTS			OUTPUT
OE	Clock	D	
L	R	H	H
L	R	L	L
H	X	X	No change

82 74HC386 Quad 2-Input Exclusive OR Gate

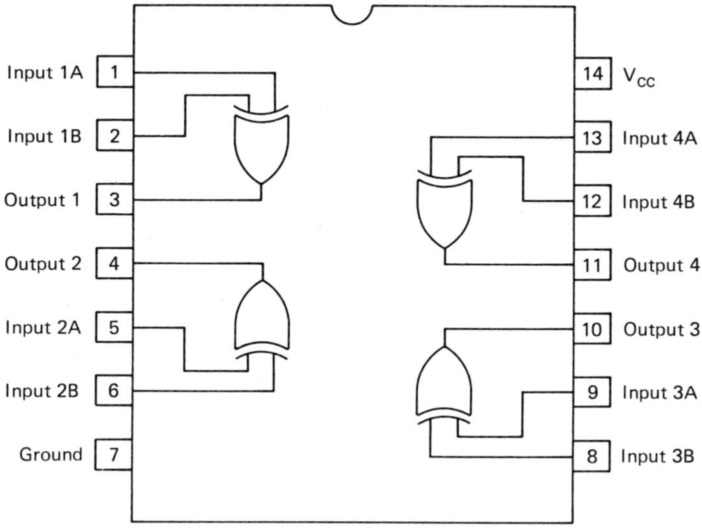

This device contains four independent two-input exclusive OR (XOR) gates. It is identical to the 74HC86 but has different pin assignments. Maximum propagation delay is 120 ns at 2 V and 24 ns at 5 V. Quiescent mode current consumption is 2 μA per package. The logic table for each gate is:

INPUTS		OUTPUT
A	B	
L	L	L
L	H	H
H	L	H
H	H	L

83 74HC390 Dual 4-Stage Binary Ripple Counter with Divide-by-Two and Divide-by-Five Sections

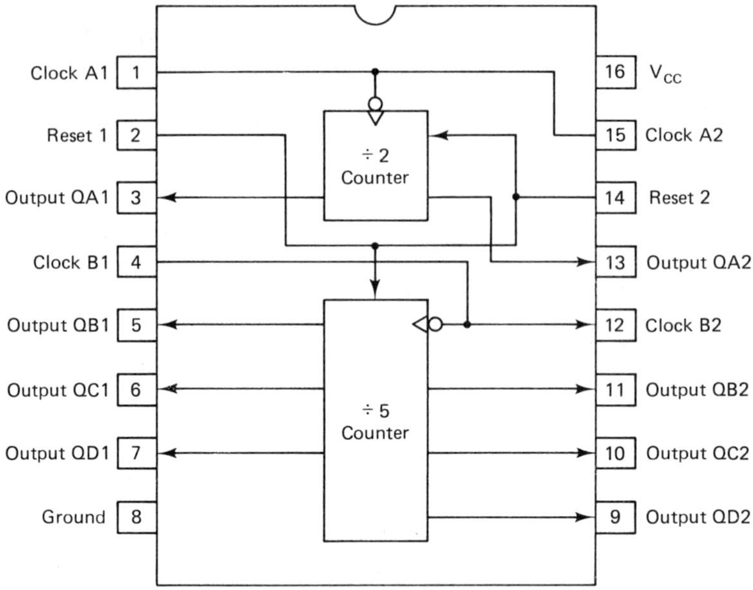

This device consists of two independent four-bit counters, each composed of a divide-by-two and a divide-by-five section. The counters have separate clock inputs and can be cascaded into divide-by-two and divide-by-five combinations up to a divide-by-100 counter. Triggering of each counter takes place on the negative edge of the clock signal. Each counter has a separate asynchronous input that must be kept low for normal operation. Maximum clock frequency is 5.4 MHz at 2 V and 27 MHz at 5 V. Maximum propagation delay is 120 ns at 2 V and 24 ns at 5 V. The function table for each counter section is:

CLOCK A	CLOCK B	RESET	RESULT
X	X	H	Reset both counters
F	X	L	Divide-by-two operation
X	F	L	Divide-by-five operation

84 74HC393 Dual 4-Stage Binary Ripple Counter

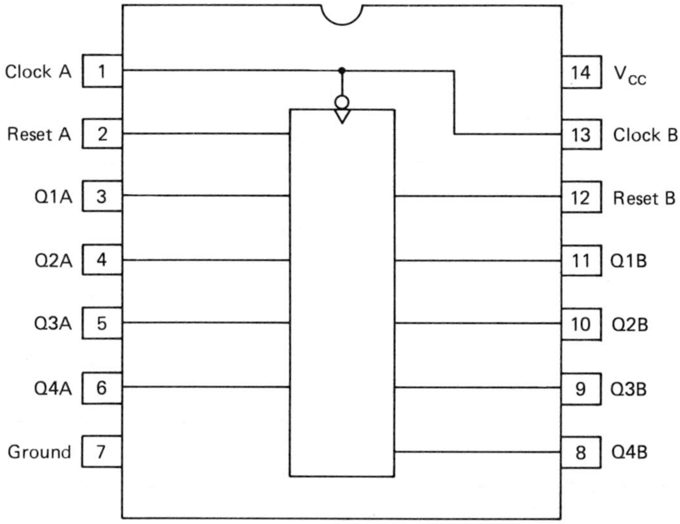

This device consists of two independent four-bit binary ripple counters with parallel outputs from each counter stage. The count advances on the negative edge of each clock pulse. There are also asynchronous result inputs at pins 2 and 12, which must be kept low for normal operation. If the reset input for a counter goes high, all outputs for that counter are made low. Two of these devices can be cascaded to produce a divide-by-256 counter. The maximum clock frequency is 5.4 MHz at 2 V and 27 MHz at 5 V. The function table for each counter is:

CLOCK	RESET	OUTPUTS
X	H	All low
H	L	No change
L	L	No change
R	L	No change
F	L	Advance one count

85 74HC533 Octal 3-State Inverting Transparent Latch

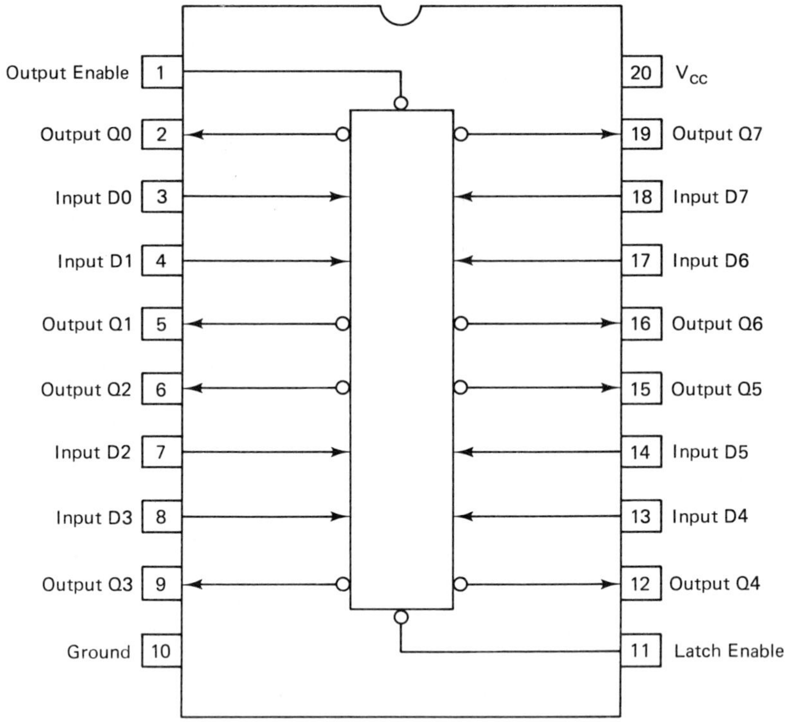

This device is similar in function to the 74HC373, but it has inverting outputs. There are eight data inputs, labeled D0 through D7, and eight inverting data outputs, labeled Q0 through Q7. When the latch enable (LE) input is high, the signals present at the data inputs appear at the outputs in inverting form. If the latch enable is low, the outputs will not change regardless of any changes in the inputs. For normal operation, the output enable (OE) input must be kept low; if it is high, the outputs will be set to a high impedance. A high output enable input will not affect any of the data stored in the latch, and the latched outputs can be restored by returning the output enable signal to low. This device is available in an HCT version. Maximum propagation delay is 150 ns at 2 V and 30 ns at 5 V. The function table for this device is:

INPUTS			OUTPUT
OE	LE	D	
L	H	H	L
L	H	L	H
L	L	X	No change
H	X	X	Z

86 74HC534 Octal 3-State Inverting D Flip-Flop

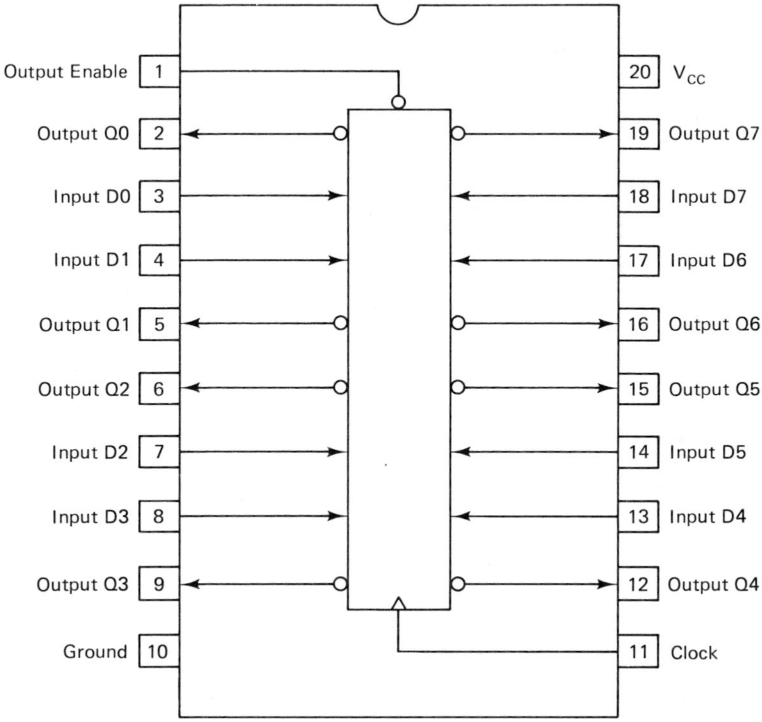

This device is similar to the 74HC374 but it has inverting outputs. It consists of eight D flip-flops sharing common clock and output enable (OE) inputs. Data present at inputs D0 through D7 are transferred to the corresponding outputs, labeled Q0 through Q7, on the positive edge of each clock pulse. The logic levels at the inputs are inverted at the outputs. For normal operation, the output enable must be held low. If the output enable is high, the outputs have a high impedance. A high output enable does not affect the state of the flip-flops. This device is available in an HCT version. Maximum clock frequency is 6 MHz at 2 V and 30 MHz at 5 V. The function table for this device is:

INPUTS			OUTPUT
OE	Clock	D	
L	R	H	L
L	R	L	H
L	L	X	No change
L	H	X	No change
L	F	X	No change
H	X	X	Z

87 74HC540 Octal 3-State Inverting Buffer/Line Driver/Line Receiver

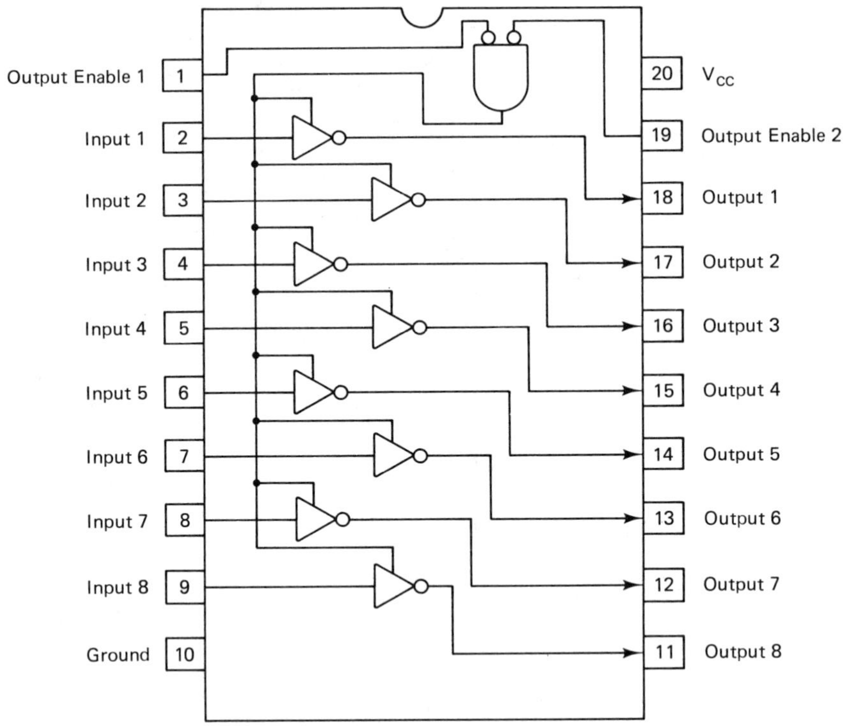

This device consists of eight separate inverting buffers sharing two output enable inputs (labeled OE1 and OE2), and it is designed for use in bus-oriented systems. Both output enable inputs must be at a low logic level for normal operation. If at least one output enable is high, the outputs are at a high impedance. This device is available in an HCT version. Maximum propagation delay is 125 ns at 2 V and 25 ns at 5 V. The logic table for each buffer is:

INPUTS			OUTPUT
OE1	OE2	Input	
L	L	L	H
L	L	H	L
H	X	X	Z
X	H	X	Z

88 74HC541 Octal 3-State Noninverting Buffer/Line Driver/Line Receiver

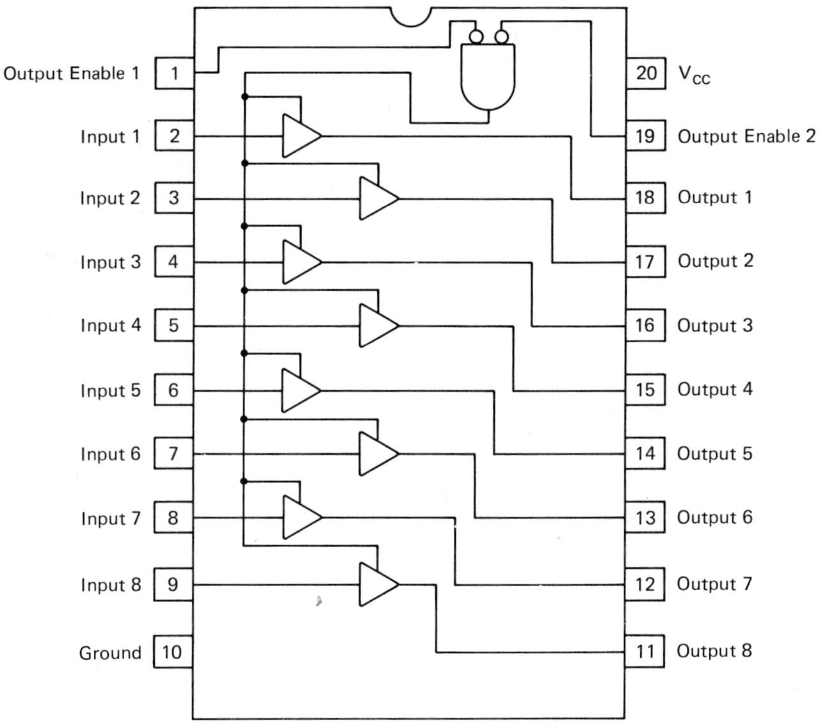

This device is identical to the 74HC540 but has noninverting outputs. It is also available in an HCT version. The logic table for each buffer is:

INPUTS			OUTPUT
OE1	OE2	Input	
L	L	L	L
L	L	H	H
H	X	X	Z
X	H	X	Z

89 74HC563 Octal 3-State Inverting Transparent Latch

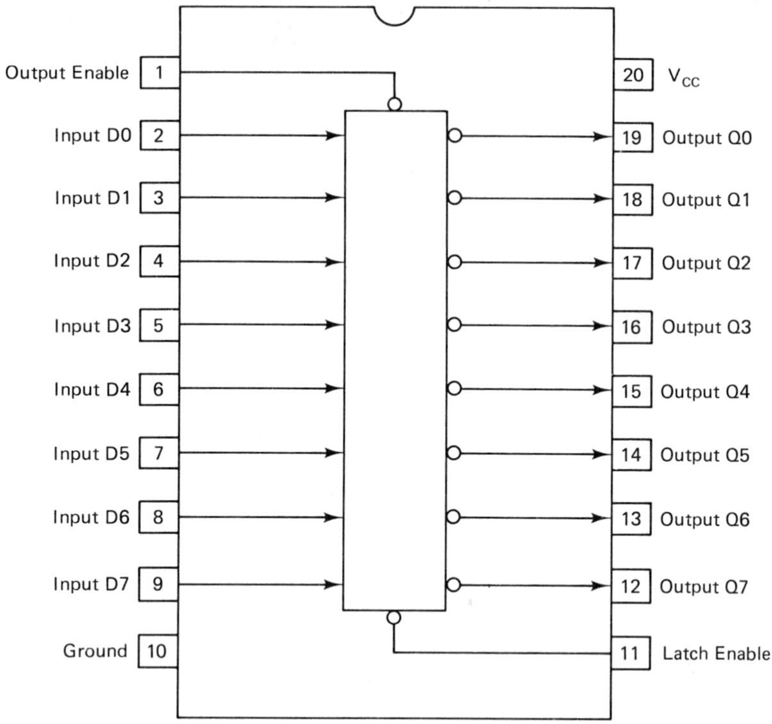

This device is identical in function to the 74HC533, but the pin arrangement has been altered so that the data inputs are on the side of the IC package opposite the outputs. For operational information on this device, refer to the details given for the 74HC533.

90 74HC564 Octal 3-State Inverting D Flip-Flop

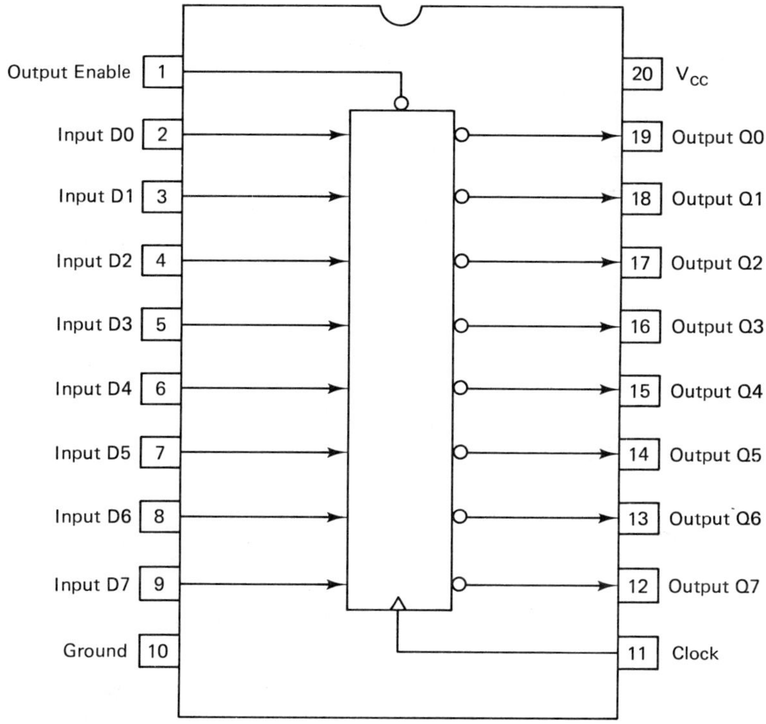

This device is identical to the 74HC534 but has a different pin arrangement so that all flip-flop inputs are on the side of the IC package opposite the outputs. For operational information on this device, refer to the details given for the 74HC534.

91 74HC573 Octal 3-State Noninverting Transparent Latch

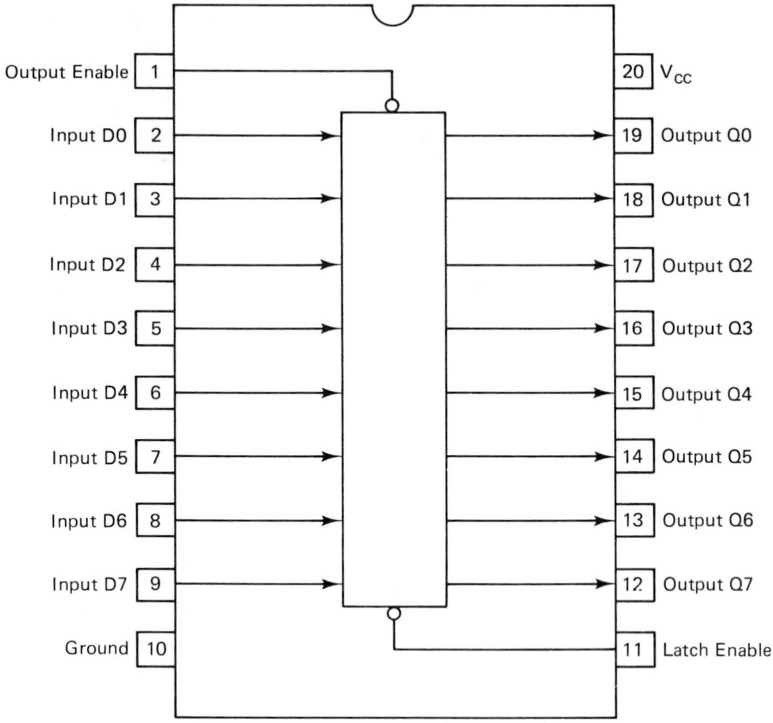

This device is identical to the 74HC373 but has a different pin arrangement so that all data inputs are on the side of the IC package opposite the outputs. For operational information on this device, refer to the details given for the 74HC373.

92 74HC574 Octal 3-State Noninverting D Flip-Flop

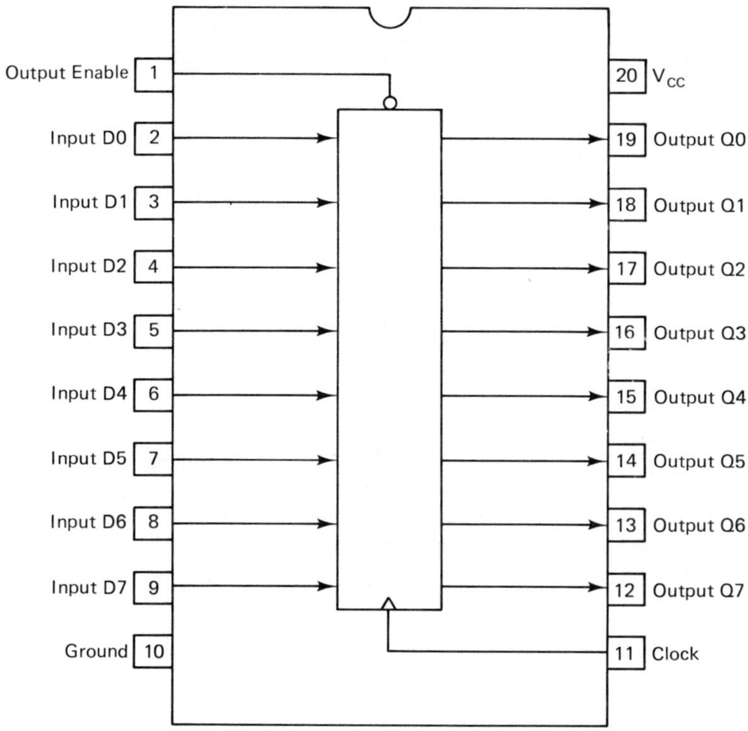

This device is identical to the 74HC374 but has a different pin arrangement to place the flip-flop inputs on the side of the package opposite the outputs. For operational information on this device, refer to the details given for the 74HC374.

93 74HC589 8-Bit Serial or Parallel Input/Serial Output Shift Register with 3-State Output

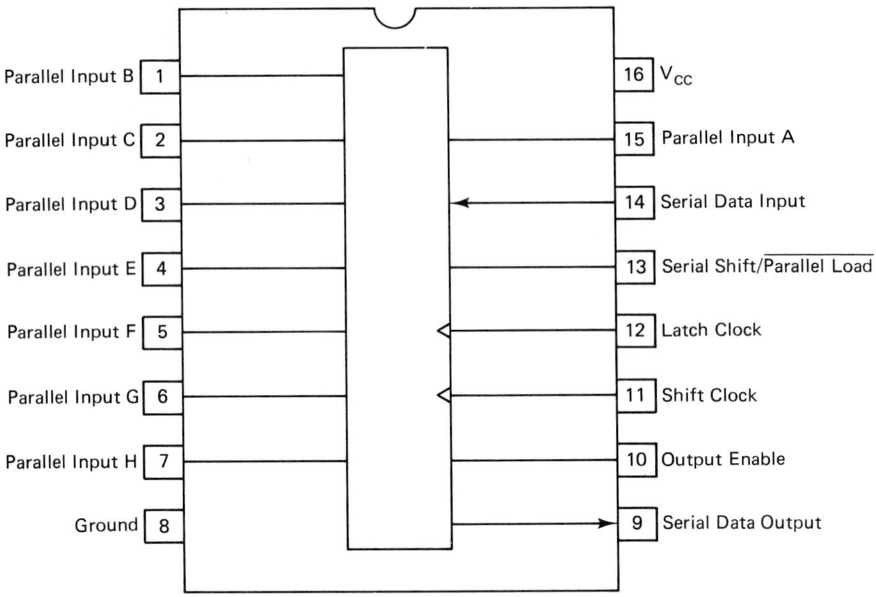

This device consists of an eight-bit internal storage latch that feeds parallel data into an eight-bit shift register via eight parallel inputs labeled A through H. Data can also be loaded serially directly into the shift register via a serial data (SD) input at pin 14. There are two separate clock inputs, the latch clock (LC) and shift clock (SC), and the device responds to positive edges of both clock signals. The latch clock signal latches the shift register data, while the shift clock signal causes data at the serial data input to be loaded into the shift register. There is a serial shift/parallel load input at pin 10, which should be held high for normal operation. A low input at this pin sets the shift register contents to low but does not affect the eight-bit latch contents. There is an output enable (OE) input at pin 10, which should be kept low for normal operation; when it is high, all outputs have a high impedance. Maximum clock frequency is 6 MHz at 2 V and 30 MHz at 5 V. In the following table, "NP" under the heading for a clock signal indicates that the clock input can be anything but the positive edge of a clock signal—a negative edge, a steady high, or a steady low state. "D" indicates serial input data. The function table for this device is:

INPUTS					OPERATION
Reset	SD	SC	LC	OE	
L	X	X	NP	L	Reset shift register
H	D	R	NP	L	Shift serial data into shift register
H	X	NP	NP	L	Shift register unchanged
H	X	NP	R	L	Transfer register contents to latch
X	X	X	NP	L	Latch unchanged
X	X	X	X	L	Enable parallel outputs
X	X	X	X	H	Outputs show high impedance

94 74HC595 8-Bit Serial Input/Serial or Parallel Output Shift Register with Latched 3-State Outputs

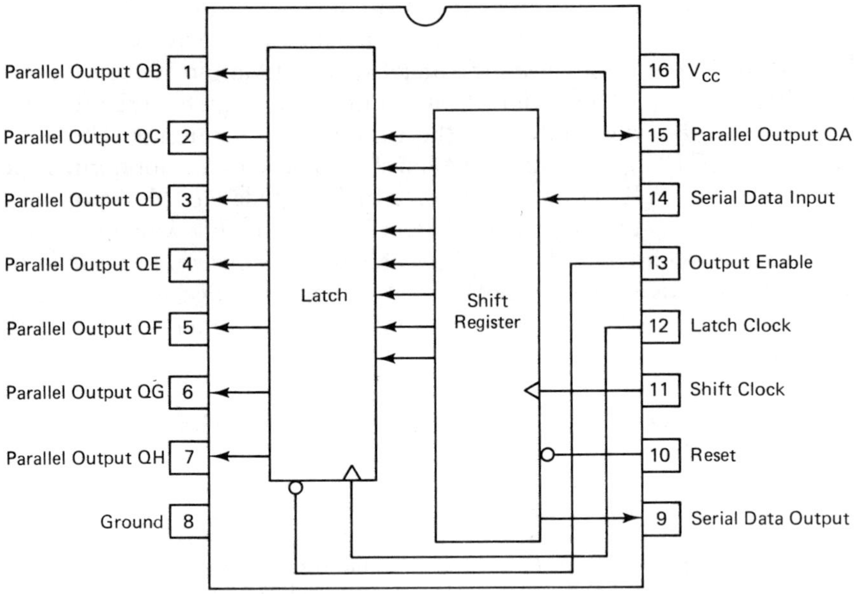

This device consists of an eight-bit serial shift register and an eight-bit D-type latch with three state parallel outputs labeled QA through QH. The shift register accepts serial data at the serial data (SD) input at pin 14 and provides a serial output (SO) at pin 9. There are separate shift clock (SC) and latch clock (LC) inputs. The shift clock input is fed to the shift register, while the latch clock controls the latch; both respond to the positive edge of each clock signal. There is also an asynchronous reset (RST) input for the shift register (pin 10), which should normally be kept high for normal operation; setting it to low causes the contents of the shift register to be set to low. There is an output enable (OE) input, which should be kept low for normal operation. When output enable is high, the outputs all have a high impedance (although the contents of the shift register and latch are unaffected). The maximum clock signal frequency is 6 MHz at 2 V and 30 MHz at 5 V. Maximum quiescent mode power consumption is 8 μA. In the following table, "D" represents the signal at the serial data input and "NP" denotes any clock signal other than the positive edge.

INPUTS					OPERATION
RST	SD	SC	LC	OE	
L	X	X	NP	L	Reset shift register
H	D	R	NP	L	Load data into shift register
H	X	NP	NP	L	Shift register unchanged
H	X	NP	R	L	Transfer shift register contents to latch
X	X	X	NP	L	Latch contents unchanged
X	X	X	X	L	Enable parallel outputs
X	X	X	X	H	Outputs at high impedance

95 74HC597 8-Bit Serial or Parallel Input/Serial Output Shift Register with Input Latch

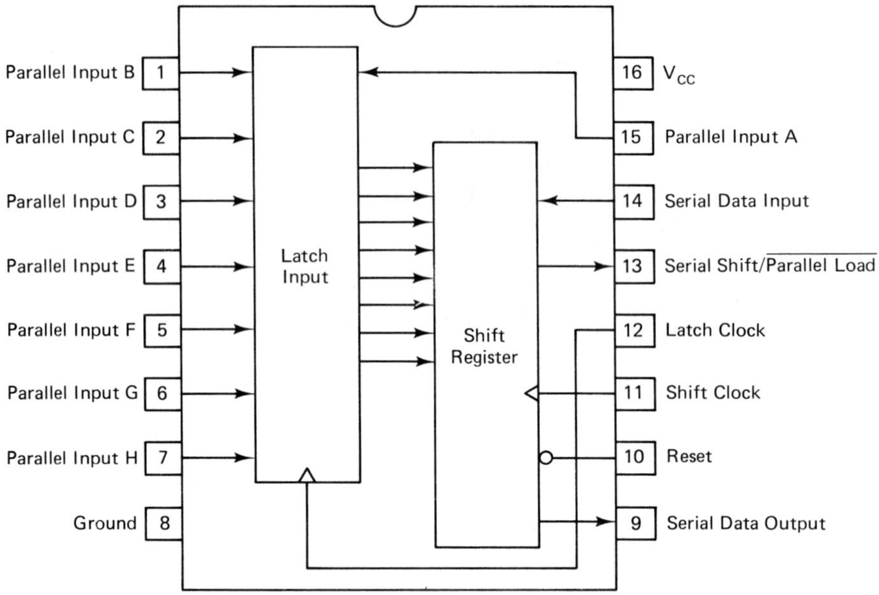

This device is similar in function to the 74HC589, but it does not have an output that can be placed at a high impedance. There are eight parallel inputs, labeled A through H, which feed into a latch section. The outputs of the latch are fed into a shift register. There is a serial data (SD) input at pin 14, which is fed directly into the shift register. There are latch clock (LC) and shift clock (SC) inputs; these sections operate on the positive edge of a clock pulse. There is a serial shift/parallel load (SS/PL) input at pin 13, which is used to control the loading and shifting of data. The reset (RST) input at pin 10 should be kept high for normal operation; if it is low, the contents of the shift register are set to low. The maximum clock frequency is 6 MHz at 2 V and 30 MHz at 5 V. In the following function table, "D" represents the serial data input signal while "NP" represents any clock signal other than a positive-edge one.

INPUTS					OPERATION
RST	SS/PL	LC	SC	SD	
L	X	NP	X	X	Reset shift register
L	X	R	X	X	Reset shift together and load parallel inputs
H	H	R	NP	X	Load parallel data into latch
H	L	NP	X	X	Transfer latch contents to shift register
H	H	NP	NP	X	Contents of latch and register unchanged
H	L	R	X	X	Load parallel data into latch and shift register
H	H	X	R	D	Load serial data
H	H	R	R	D	Load parallel data into latch and shift serial data into shift register

96 74HC640 Octal 3-State Inverting Bus Transceiver

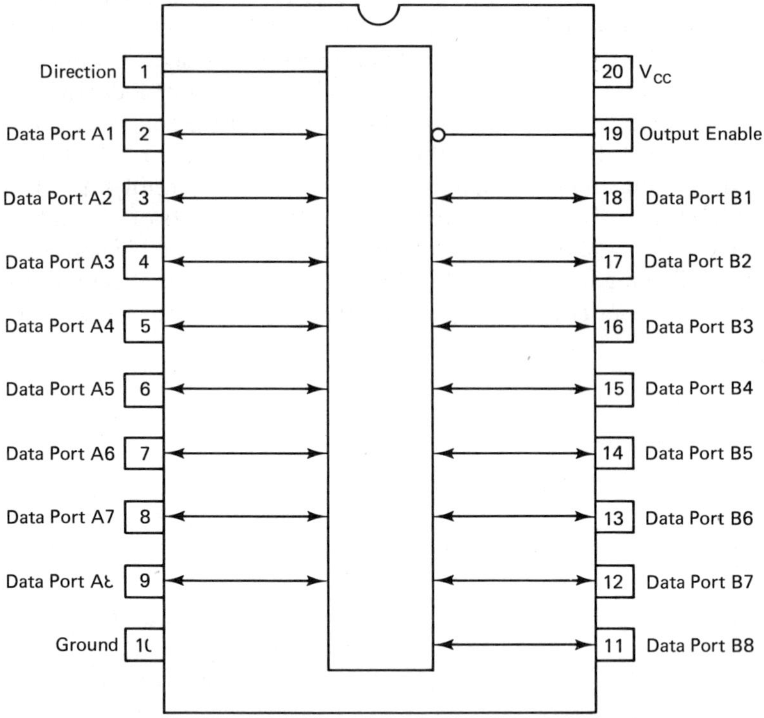

This device is used for two-way asynchronous communication between eight-bit data buses, performing functions similar to the 74HC245 and 74HC643 devices. There are two separate data ports, labeled A and B, which each consist of eight parallel inputs/outputs. The direction of data transfer is controlled by the direction input (pin 1). A low direction input signal causes data to flow from B to A, while a high direction input signal causes data to flow from A to B. The output data are inverted from their input states. There is an output enable (OE) input at pin 19, which should be kept low for normal operation; when it is high, all data ports have high impedance and the buses are isolated from each other. This device is available in an HCT version. Maximum propagation delay is 100 ns at 2 V and 20 ns at 5 V. Quiescent mode current consumption is 8 μA. The function table for this device is:

INPUTS		OPERATION
OE	Direction	
L	L	Data move from bus B to bus A
L	H	Data move from bus A to bus B
H	X	Buses isolated; all ports high impedance

97 74HC643 Octal 3-State Inverting and Noninverting Bus Transceiver

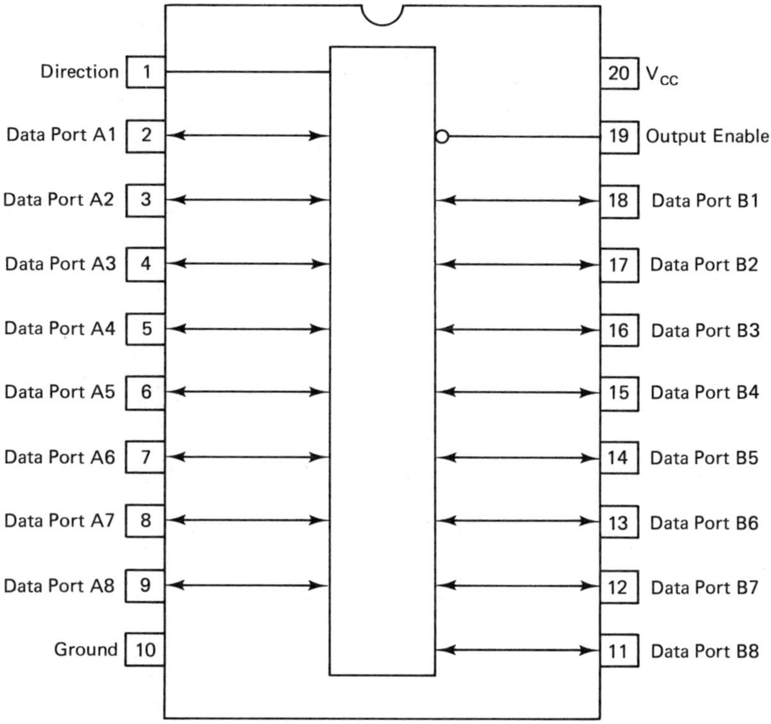

This device is used for two-way asynchronous communications between eight-bit data buses labeled A and B. It performs functions similar to the 74HC245 and 74HC640 devices but inverts data moved from A to B while data moving from B to A are not inverted. The direction input (pin 1) determines the direction of the data flow. If the direction pin is low, the data move from B to A; if the pin is high, the data move from A to B and appear at the output in inverted form. The output enable (OE) input must be low for normal operation. If it is high, the data ports are at a high impedance and the data buses are isolated from each other. This device is available in an HCT version. Maximum propagation delay is 100 ns at 2 V and 20 ns at 5 V. The function table for this device is:

INPUTS		OPERATION
OE	Direction	
L	L	Data move from bus B to bus A
L	H	Data move from bus A to bus B (inverted output)
H	X	Buses isolated; all ports high impedance

98 74HC646 Octal 3-State Bus Transceiver and D Flip-Flops with Noninverting Outputs

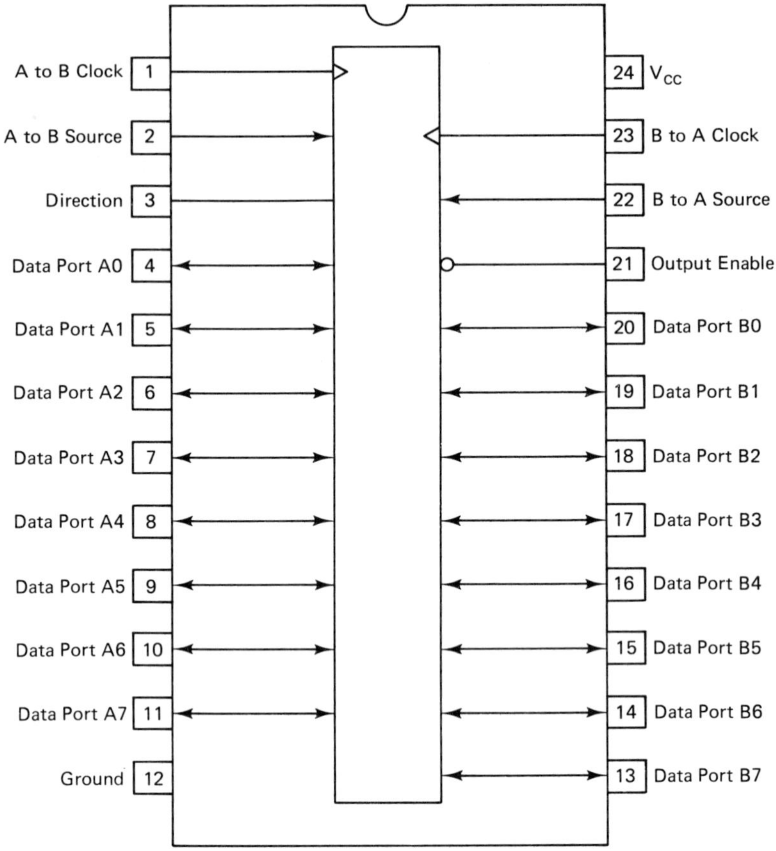

This device permits two-way communication between eight-bit buses in either asynchronous or clocked modes using internal D flip-flops. There are two sets of eight inputs/outputs, labeled A and B, which can be used for communication between data buses. In addition, these two sets of "data ports" can simultaneously be used as inputs for the internal D flip-flops. The internal flip-flops come in two groups labeled A and B, each set corresponding to the data ports. The internal D flip-flops can also send their outputs to the A and B "ports" simultaneously. The direction (DIR) input at pin 3 controls the direction of the communication between the two buses. If it is high, the data transfer is from bus A to bus B; if it is low, the direction is from bus B to bus A. Two "source" inputs at pins 2 and 22 control whether bus communication is asynchronous or clocked through the internal flip-flops. These are labeled as A to B source (ABS) and B to A source (BAS). In the asynchronous mode, the status of the internal flip-flops is not affected by any of the input or output signals. If a clocked mode of operation is used, an A to B clock (ABC) or B to A clock (BAC) signal, as appropriate, must be provided. The positive edge of each clock pulse serves as the clock signal. There is an output enable (OE) input at pin 21 which must be kept low for normal operation; if it is high, all input/output "ports" are at a high impedance and communication between the buses is impossible. However, inputs may be clocked into the internal flip-flops using the ports if appropriate clock signals are present. Maximum clock frequency is 6 MHz at 2 V and 30 MHz at 5 V. The maximum propagation delay between buses in the asynchronous mode is 170 ns at 2 V and 34 ns at 5 V. In the following function table, "NP" denotes any clock signal or level except a positive edge.

INPUTS						OPERATION
OE	DIR	ABC	BAC	ABS	BAS	
H	X	NP	NP	X	X	All ports at high impedance
H	X	R	R	X	X	Ports may be used as flip-flop inputs
L	H	NP	X	L	X	Data move from A to B
L	H	NP	X	H	X	Output at B is status of A internal flip-flops
L	H	R	X	L	X	Data at A are clocked into internal flip-flops
L	H	R	X	H	X	Data at A are clocked into internal flip-flops and appear at B
L	L	X	NP	X	L	Data move from B to A
L	L	X	NP	X	H	Output of A is status of B internal flip-flops
L	L	X	R	X	L	Data at B are clocked into internal flip-flops
L	L	X	R	X	H	Data at B are clocked into internal flip-flops and appear at A

99 74HC648 Octal 3-State Bus Transceiver and D Flip-Flops with Inverting Outputs

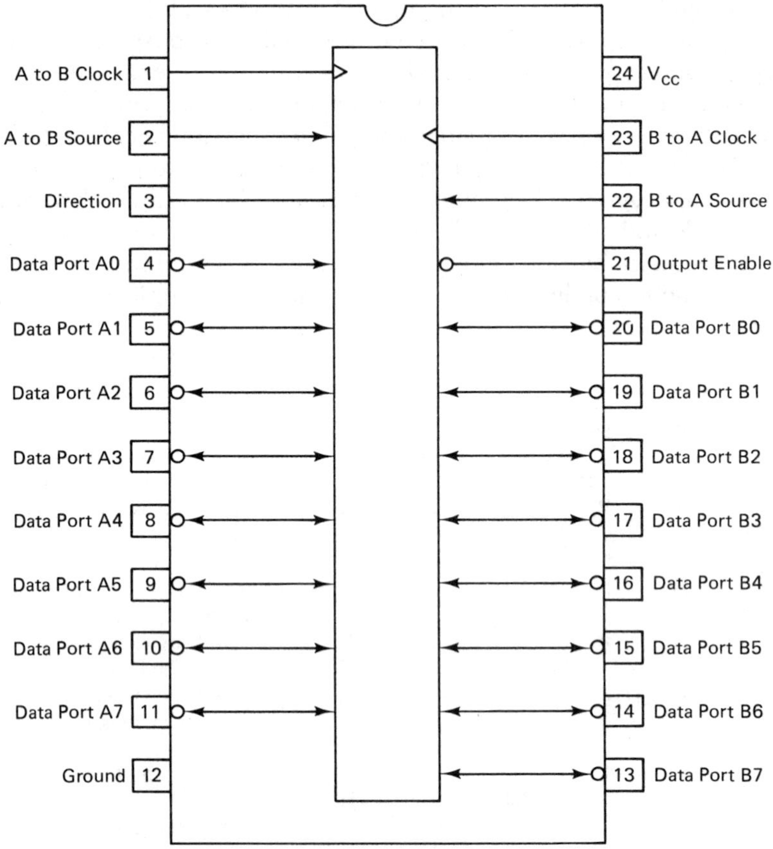

This device is identical to the 74HC646 except that the input signals are inverted at the outputs. For a description of and operating information on this device, consult the details given for the 74HC646.

100 74HC651 Octal 3-State Bus Transceiver and D Flip-Flops with Inverting Outputs

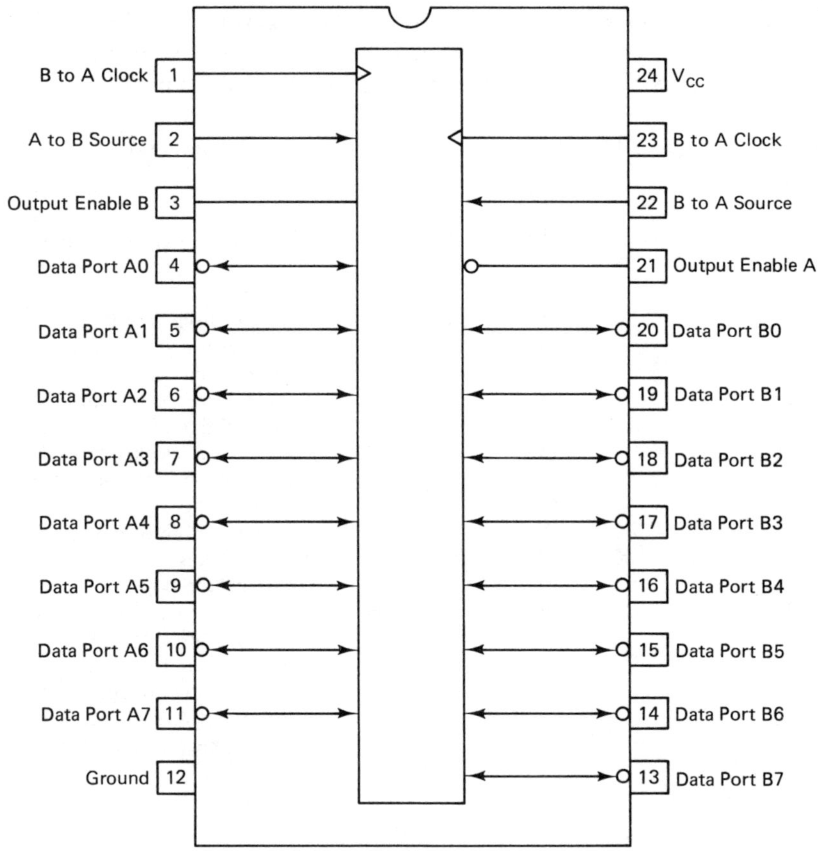

This device allows two-way communication between eight-bit buses in either asynchronous or clocked modes using internal D flip-flops. The input signals appear at the outputs in inverted form. There are two sets of inputs/outputs, labeled A and B, which are known as "data ports." There are two "source" inputs that determine whether communication between buses is asynchronous or clocked through the flip-flops. These are the A to B source (ABS) input at pin 2 and the B to A source (BAS) input at pin 22. If a clocked mode is selected, the operation is controlled by an A to B clock (ABC) or a B to A clock (BAC) input. Each positive edge of a clock pulse represents a clock signal. There are also two output enable inputs, labeled output enable A (OEA) and output enable B (OEB). The input/output "ports" can be set to a high impedance by the signals applied to these two inputs. If either or both of the sets of data ports are set to high impedance, the inputs/outputs can still be used to input data directly into the internal flip-flops; data may be clocked into the internal flip-flops at any time. Both the A and B "ports" may be used simultaneously to input data to the flip-flops. Maximum clock frequency is 6 MHz at 2 V and 30 MHz at 5 V. Maximum propagation delay between buses in the asynchronous mode is 170 ns at 2 V and 34 ns at 5 V. In the following function table, "NP" denotes any clock signal other than a positive-edge one.

INPUTS						OUTPUTS
OEB	OEA	ABC	BAC	ABS	BAS	
L	H	NP	NP	X	X	Data ports at high impedance
L	H	R	R	X	X	Ports may be used as flip-flop inputs
H	H	X	X	L	X	Data flow from A to B
H	H	X	X	H	X	Output at B is status of A internal flip-flops
H	H	R	X	L	X	Data at A are clocked into internal flip-flops
H	H	R	X	H	X	Data at A are clocked into internal flip-flops and appear at B
L	L	X	X	X	L	Data move from B to A
L	L	X	R	X	L	Output at A is status of B internal flip-flops
L	L	X	R	X	H	Data at B are clocked into internal flip-flops and appear at A
H	L	X	X	H	H	Data in A flip-flops appear at B outputs, and data in B flip-flops appear at A outputs

101 74HC652 Octal 3-State Bus Transceiver and D Flip-Flops with Noninverting Outputs

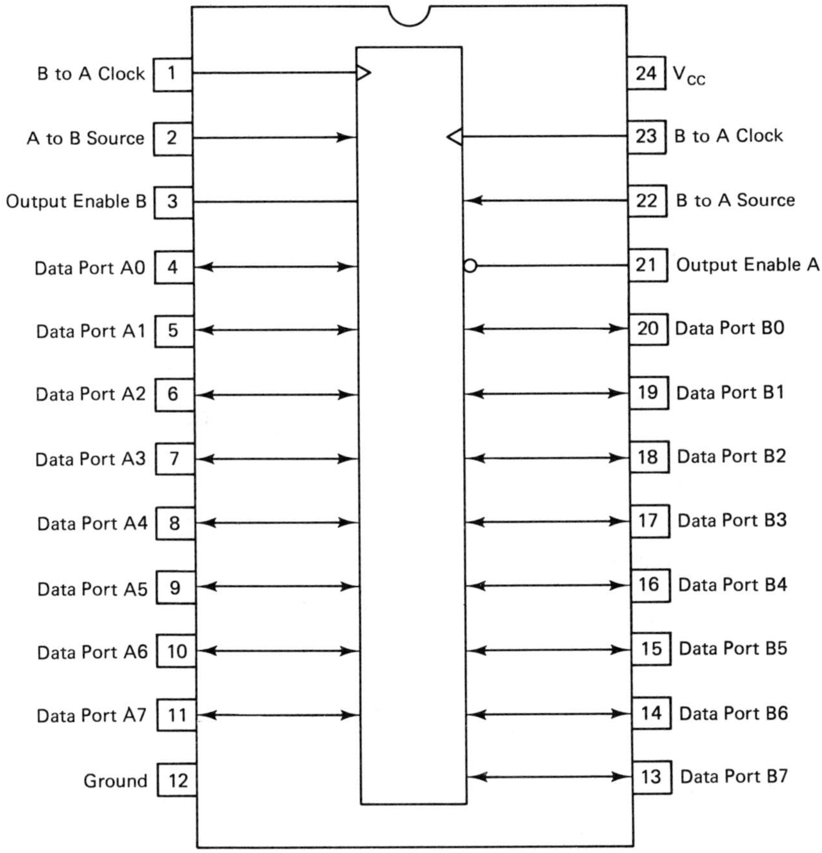

This device is identical to the 74HC651 except that the inputs do not appear at the outputs in inverted form. For details on this device and its operation, consult the details given for the 74HC651.

102 74HC688 8-Bit Equality Comparator

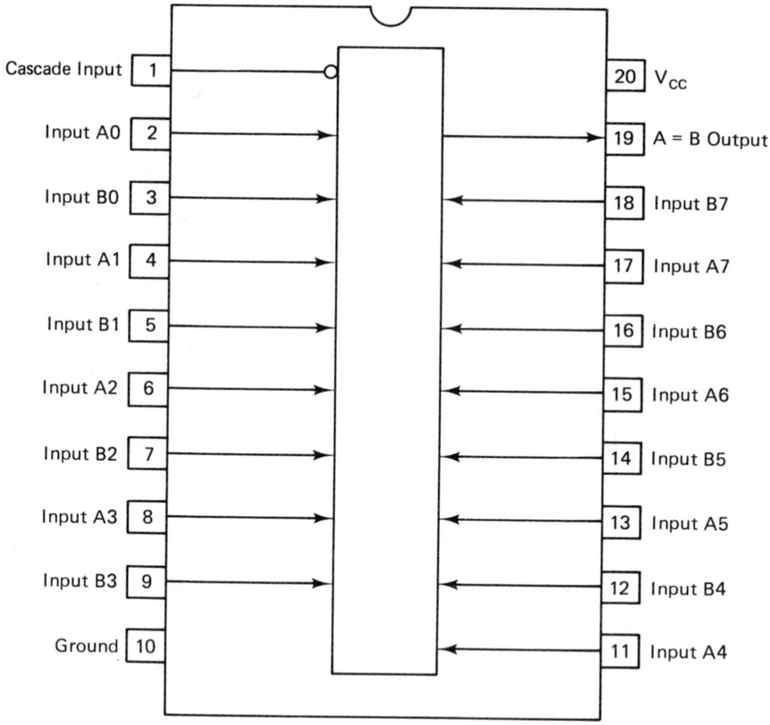

This device compares two eight-bit binary or BCD words and indicates whether they are equal. Inputs for the two words are labeled as A and B, and there is a single A = B output (pin 19). If the words are equal, the output will be low; if they are not equal, the output will be high. There is also a cascade input (pin 1), which can be used to cascade devices to compare words longer than eight bits. It should be kept low for normal operation if only one device is used. Leaving it at a high level will produce a high output from pin 19 regardless of the data words being compared. Maximum propagation delay is 210 ns at 2 V and 42 ns at 5 V. Maximum quiescent mode current consumption is 8 μA. or less.

INPUTS		OUTPUT
Words	Cascade	
A = B	L	L
A > B	L	H
A < B	L	H
A > B	L	H
X	H	H

103 74HC4002 Dual 4-Input NOR Gate

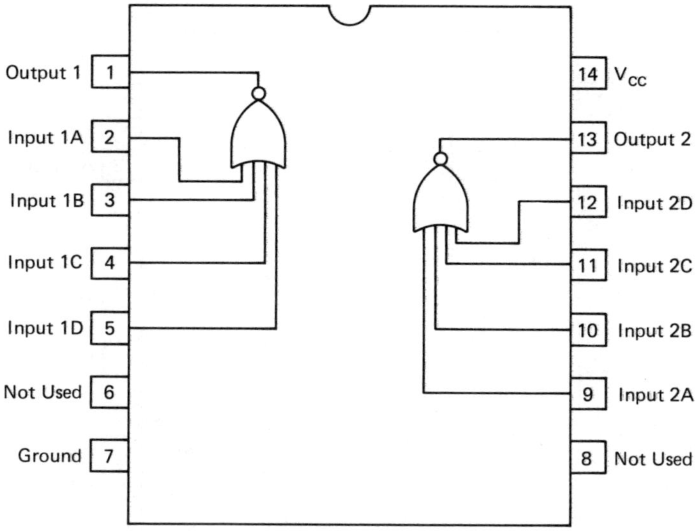

This device is the high-speed CMOS functional equivalent of the 4002 standard CMOS device. It consists of two separate NOR gates with four inputs. If all the inputs are low, the output of the gate will be high. If one or more of the inputs are high, the output will be low. Maximum propagation delay is 120 ns at 2 V and 24 ns at 5 V. Maximum quiescent supply current per package is 2 μA. The logic table for each gate is:

INPUTS				OUTPUT
A	B	C	D	
L	L	L	L	H
H	X	X	X	L
X	H	X	X	L
X	X	H	X	L
X	X	X	H	L

104 74HC4016 Quad Analog Switch/Multiplexer/Demultiplexer

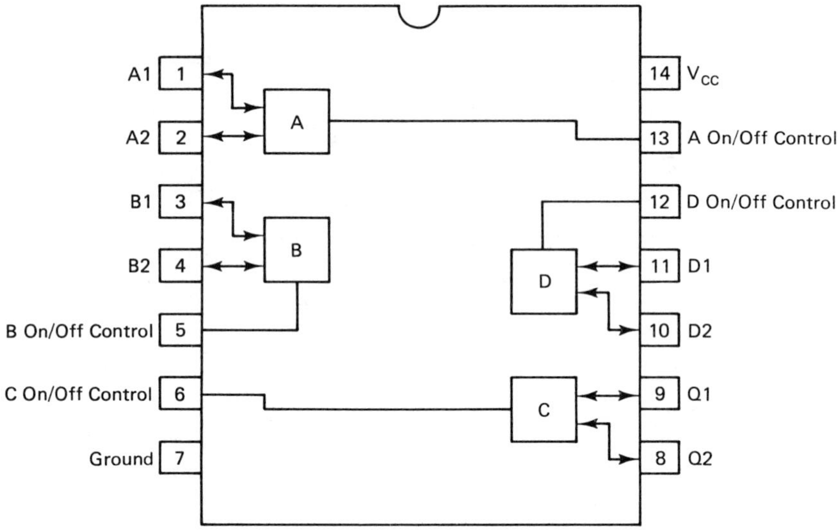

This device is the high-speed CMOS functional equivalent of the 4016 standard CMOS device. It consists of four separate analog switches controlled by corresponding on/off input signals. A low signal to an on/off input will turn the switch off, while a high signal will turn the switch on. Each switch is capable of switching analog or digital voltage levels. The supply voltage for this device can range from +2 to +12 V, and the switches can handle any voltage level equal to or less than the supply voltage used. Maximum quiescent mode supply current is 2 μA at 6 V and 8 μA at 12 V.

105 74HC4017 Decade Counter

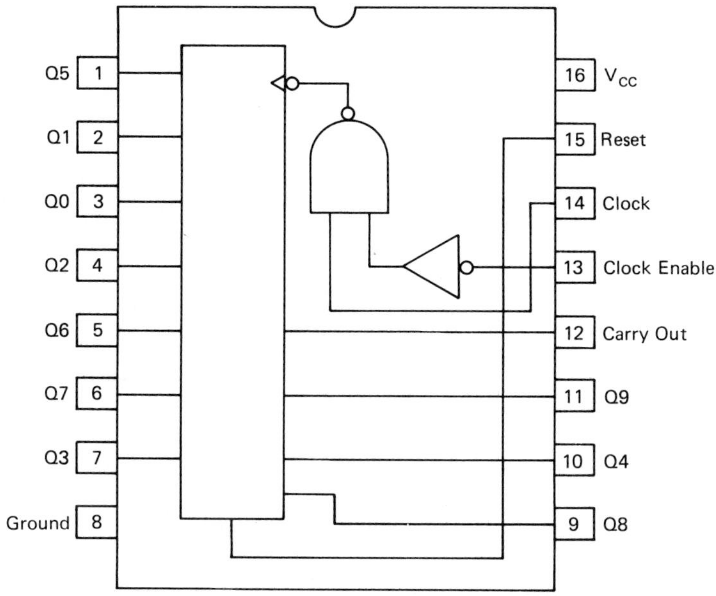

This device is the high-speed CMOS functional equivalent of the 4017 standard CMOS device. It advances a count one step with each positive edge of a clock input. The output of the counter is ten outputs, labeled Q0 through Q9, along with a separate carry output at pin 12 for overflow and cascading counters. There is a "normal" positive-edge clock input (CLK) at pin 14 and a clock enable (CE) input that can serve as a negative-edge clock input if pin 14 is held to a high logic level. The reset (RST) input at pin 15 should be held to a low level for normal operation. If it is made high, Q0 is high and outputs Q1 through Q9 are all low; the carry output is also high. The maximum clock frequency is 4 MHz at 2 V and 20 MHz at 5 V. Propagation delay from clock input to counter outputs is 230 ns at 2 V and 46 ns at 5 V. Maximum quiescent mode current consumption is 8 μA. The function table for this device is:

INPUTS			OPERATION
CLK	CE	RST	
L	X	L	No change
X	H	L	No change
X	X	H	Reset counter
R	L	L	Advance one count
F	X	L	No change
X	R	L	No change
H	F	L	Advance one count

106 74HC4020 14-Stage Binary Ripple Counter

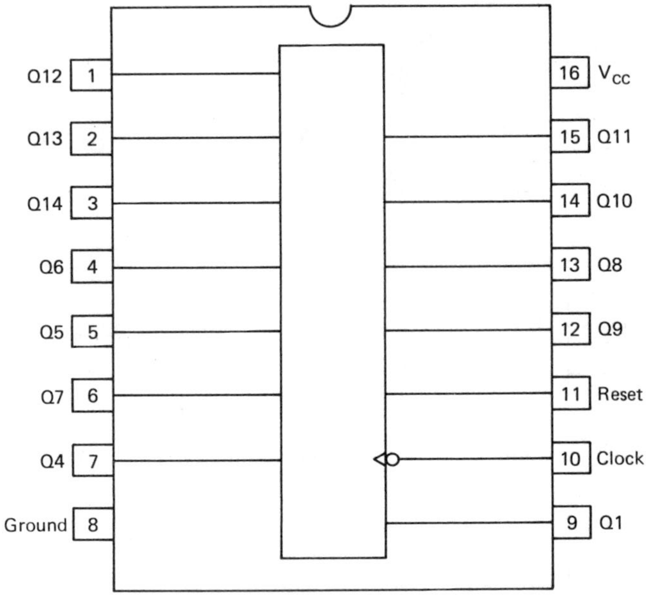

This device is the high-speed CMOS functional equivalent of the 4020 standard CMOS device. It consists of 14 master/slave flip-flops with 12 stages brought out to outputs labeled Q1 and Q4 through Q14. The output of each flip-flop feeds the next, and the frequency at each output is half that of the preceding one. The counter advances on the negative edge of each clock pulse. The reset input is asynchronous and should be kept low for normal operation; a high signal to the reset input will set all outputs to low. Due to internal ripple delays, changes in the outputs do not occur simultaneously. Maximum clock frequency is 5 MHz at 2 V and 25 MHz at 5 V. Maximum quiescent mode supply current is 8 μA. The function table for this device is:

CLOCK	RESET	OUTPUTS
R	L	No change
F	L	Advance one count
X	H	All outputs low

107 74HC4024 7-Stage Binary Ripple Counter

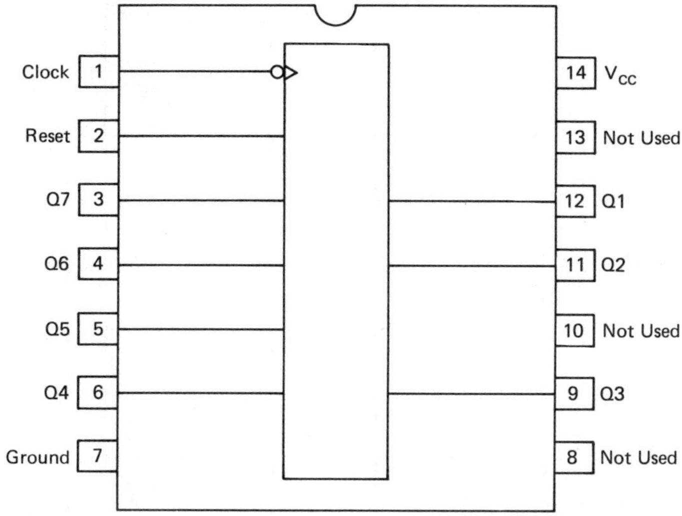

This device is the high-speed CMOS functional equivalent of the 4024 standard CMOS device. It consists of seven master/slave flip-flops with outputs labeled Q1 through Q7. The output of each flip-flop feeds the next, and the frequency at each output is half that of the preceding one. The counter advances on the negative edge of each clock pulse. The reset input is asynchronous and should be kept low for normal operation; a high input sets all outputs to low. Due to internal ripple delays, changes in the outputs do not occur simultaneously. Maximum clock frequency is 5.4 MHz at 2 V and 27 MHz at 5 V. Maximum propagation delay from the clock input to output Q1 is 210 ns at 2 V and 42 ns at 5 V. The function table for this device is:

CLOCK	RESET	OUTPUTS
R	L	No change
F	L	Advance one count
X	H	All low

108 74HC4040 12-Stage Binary Ripple Counter

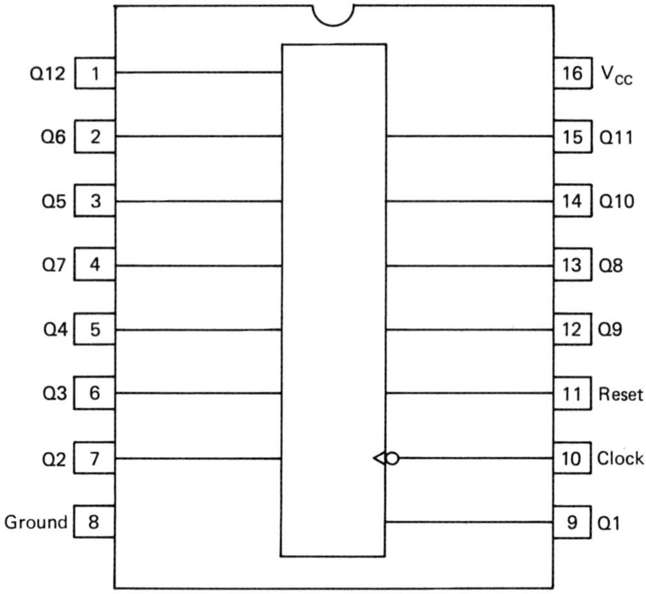

This device is the high-speed CMOS functional equivalent of the 4040 standard CMOS device. It consists of 12 master/slave flip-flops with outputs labeled Q1 through Q12. The output of each flip-flop feeds the next, and the frequency at each output is half that of the preceding one. The counter advances on each negative edge of a clock pulse. There is a reset input at pin 11, which should be kept low for normal operation; if it is high, all outputs are set to low. Due to internal ripple delays, changes in the outputs do not occur simultaneously. Maximum clock frequency is 5 MHz at 2 V and 25 MHz at 5 V. Maximum propagation delay from the clock input to Q1 is 210 ns at 2 V and 42 ns at 5 V. The function table for this device is:

CLOCK	RESET	OUTPUTS
R	L	No change
F	L	Advance one count
X	H	All low

109 74HC4046A Phase-Locked Loop with Voltage-Controlled Oscillator

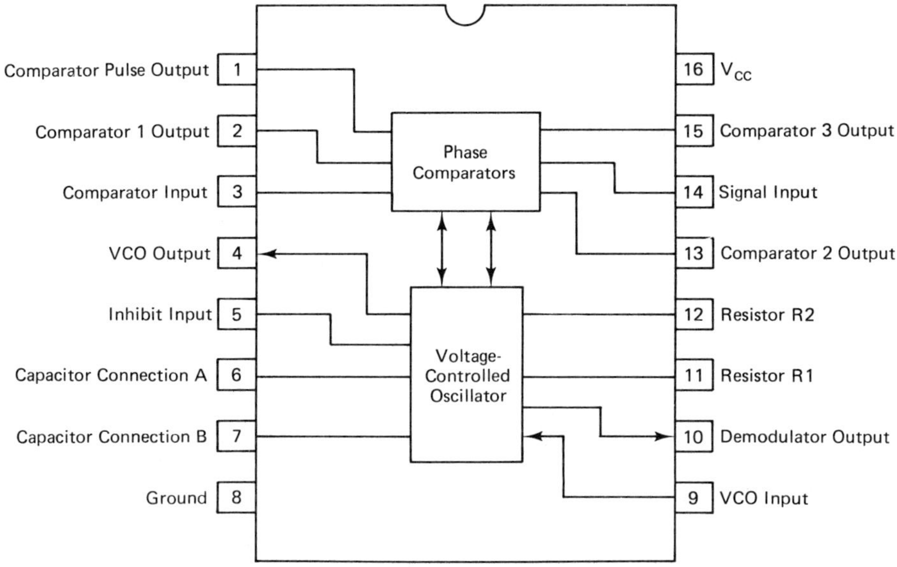

This device is the high-speed CMOS functional equivalent of the 4046A standard CMOS device. It is a phase-locked loop circuit consisting of a linear voltage-controlled oscillator (VCO) and three different phase comparators with a common signal input amplifier and a common comparator input. Comparator 1 is an exclusive OR network, comparator 2 is a positive-edge-triggered J-K flip-flop, and comparator 3 is a positive-edge-triggered R-S flip-flop. If the VCO section is used, the supply voltage must be in the range of 3 to 6 V; if the comparators alone are used, the supply voltage would be between 2 and 6 V. (This device is available in an HCT version, and its supply voltage should be restricted to 5 V.) The VCO requires one external capacitor between pins 6 and 7 and one external resistor between pin 11 and ground or two external resistors between pin 11 and ground and pin 12 and ground. Using a resistor and capacitor determines the frequency range of the VCO, while using two resistors enables the VCO to have a frequency offset if desired. Center frequency of this device is typically 17 MHz at 5 V. Quiescent mode supply current for the comparator section alone is 8 μA; when the VCO is used, this typically rises to approximately 100 μA. This device can be used in FM modulation and demodulation, frequency synthesis and multiplication, tone decoding, voltage to frequency conversion, and similar applications.

110 74HC4049 Hex Inverting Buffers/Logic-Level Down Converters

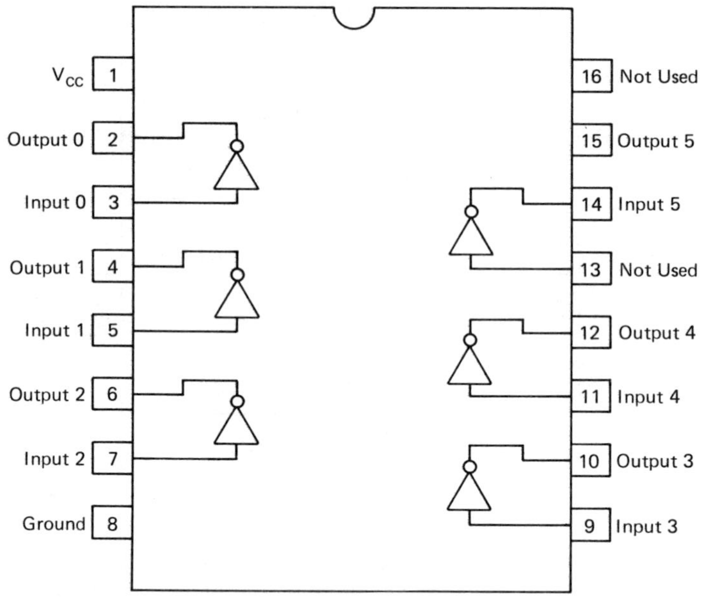

This device is the high-speed CMOS functional equivalent of the 4049 standard CMOS device. It consists of six inverting buffers in which the input protection circuitry has been modified to allow the use of input voltages up to 15 V. Thus, the buffers may be used as logic-level "translators" to convert from a high voltage to a low voltage while operating from a low-voltage power source. These devices are commonly used to interface standard CMOS to high-speed CMOS. The output of each buffer is inverted in logic level from the input.

111 74HC4050 Hex Noninverting Buffers/Logic-Level Down Converters

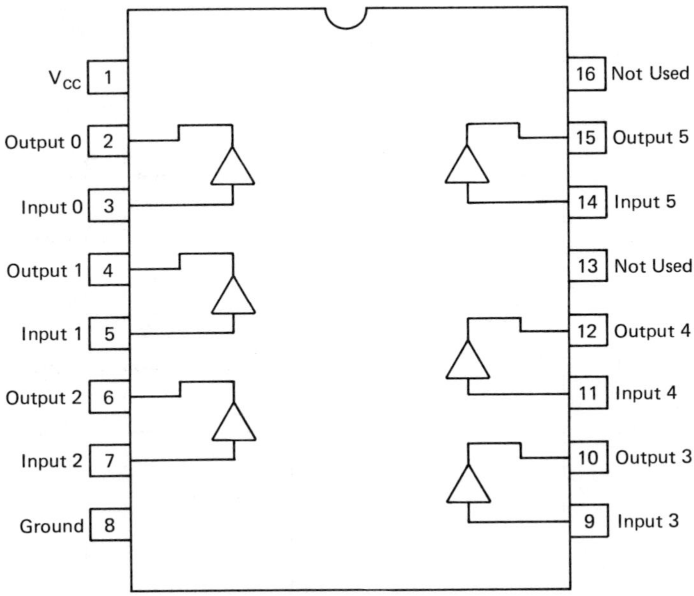

This device is the high-speed CMOS functional equivalent of the 4050 standard CMOS device. It is identical to the 74HC4049 except that the outputs of the buffers are of the same logic level as the inputs. For further information, consult the details given for the 74HC4049.

112 74HC4051 Single-Pole, 8-Position Analog Multiplexers/Demultiplexers with Common Off

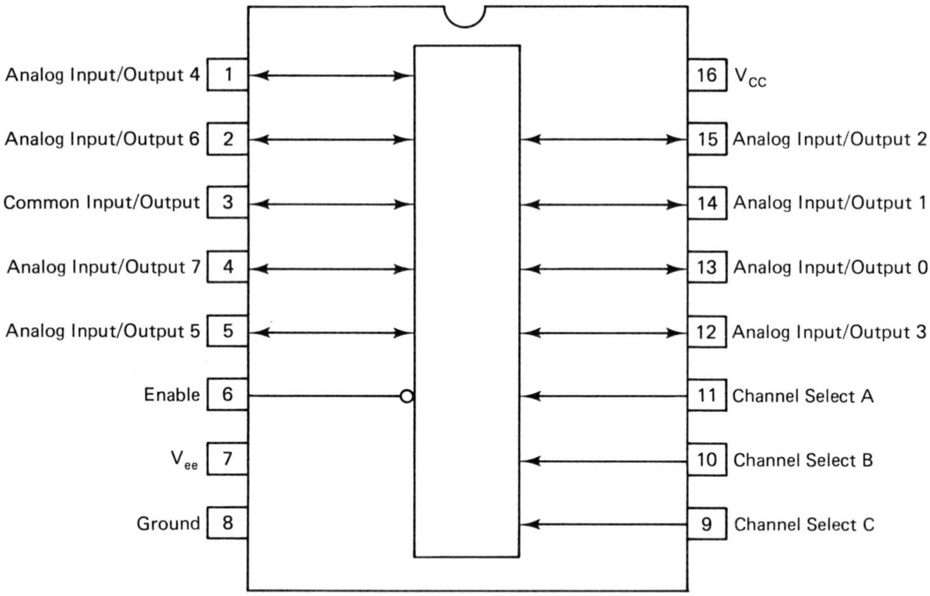

This device is the high-speed CMOS functional equivalent of the 4051 standard CMOS device. It is a switch with eight inputs/outputs labeled 0 through 7 and a single common input/output at pin 3. The inputs may be analog voltages across the complete supply voltage range. The selection of the appropriate input/output is performed by signals applied to channel select A (CSA), channel select B (CSB), and channel select C (CSC). If there is a single input signal at pin 3, these channel select inputs determine to which of the analog input/outputs the input is sent. Otherwise the channel select inputs determine which of the analog input/outputs are selected to appear at pin 3. There is an enable input at pin 6 which should be kept low for normal operation. If it is low, the switch is turned off. Vcc represents the positive supply voltage and can be from −0.5 to +7 V if referenced to ground or −0.5 to +14 V if referenced to Vee. Vee is the negative supply voltage and can be from −7 to +0.5 V if referenced to ground. Maximum quiescent mode supply current is 2 μA. The function table is:

ENABLE	CSA	CSB	CSC	INPUT/OUTPUT
L	L	L	L	0
L	L	L	H	1
L	L	H	L	2
L	L	H	H	3
L	H	L	L	4
L	H	L	H	5
L	H	H	L	6
L	H	H	H	7
H	X	X	X	None

113 74HC4052 Double-Pole, 4-Position Analog Multiplexers/Demultiplexers with Common Off

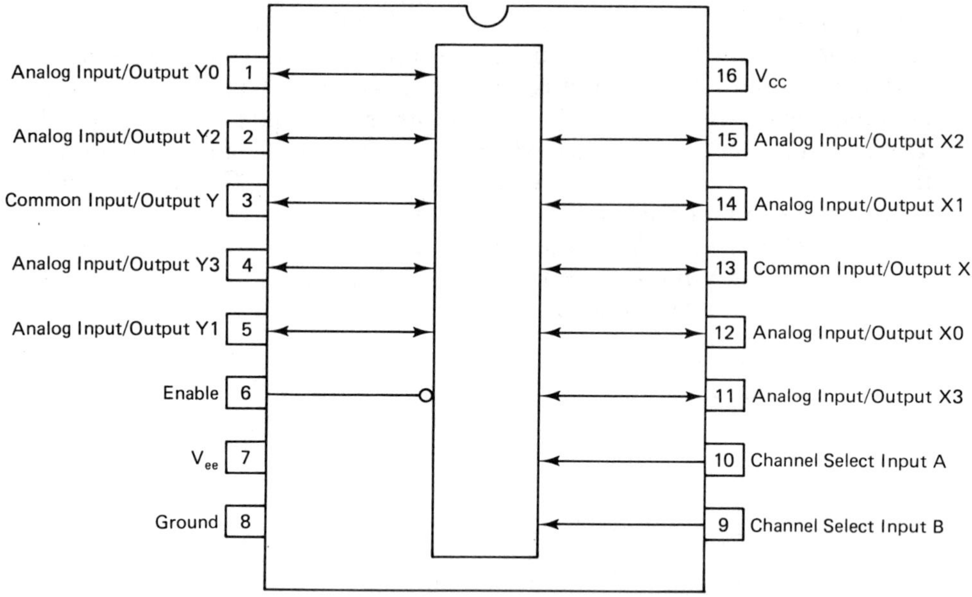

This device is the high-speed CMOS functional equivalent of the 4052 standard CMOS device. It is similar to the 74HC4051 but consists of two separate switches, each with four input/outputs numbered 0 through 3 and a common input/output. There is a common enable input at pin 6, and switching is accomplished by signals applied to channel select A (CSA) and channel select B (CSB). Electrical specifications and requirements are the same as those for the 74HC4051. The function table for each switch is:

ENABLE	CSB	CSA	INPUT/OUTPUT
L	L	L	0
L	L	H	1
L	H	L	2
L	H	H	3
H	X	X	None

114 74HC4053 Triple Single-Pole, Double-Position Analog Multiplexers/Demultiplexers with Common Off

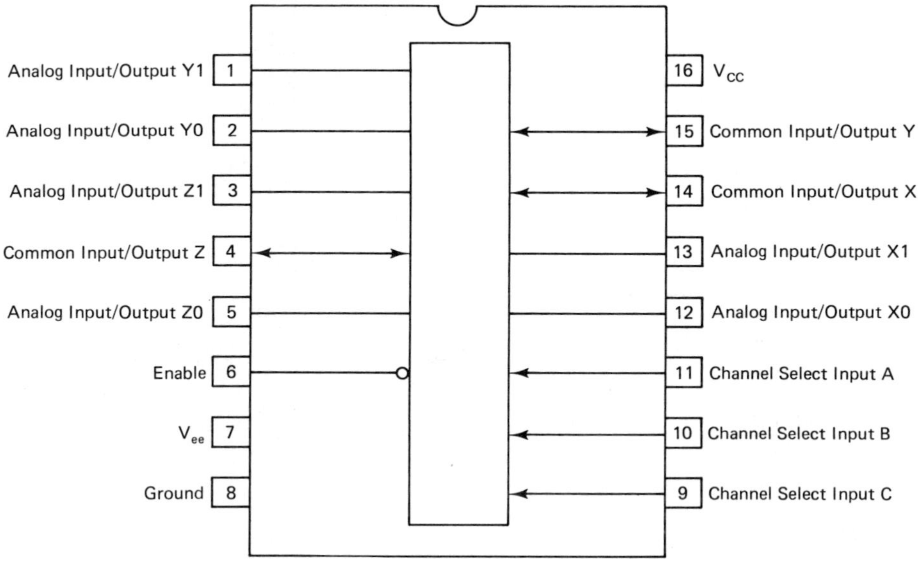

This device is the high-speed CMOS functional equivalent of the 4053 standard CMOS device. It is similar to the 74HC4051 and 74HC4052. It consists of three switches, each with a pair of analog input/outputs. These six input/outputs are labeled X0, X1, Y0, Y1, Z0, and Z1. There are common input/outputs labeled X, Y, and Z. The input/outputs are selected by signals present at channel select A (CSA), channel select B (CSB), and channel select C (CSC). The electrical specifications and requirements are the same as those for the 74HC4051 and 74HC4052. The function table for this device is:

ENABLE	CSC	CSB	CSA	INPUT/OUTPUTS		
L	L	L	L	Z0	Y0	X0
L	L	L	H	Z0	Y0	X1
L	L	H	L	Z0	Y1	X0
L	L	H	H	Z0	Y1	X1
L	H	L	L	Z1	Y0	X0
L	H	L	H	Z1	Y0	X1
L	H	H	L	Z1	Y1	X0
L	H	H	H	Z1	Y1	X1
H	X	X	X		None	

115 74HC4060 14-Stage Binary Ripple Counter with Oscillator

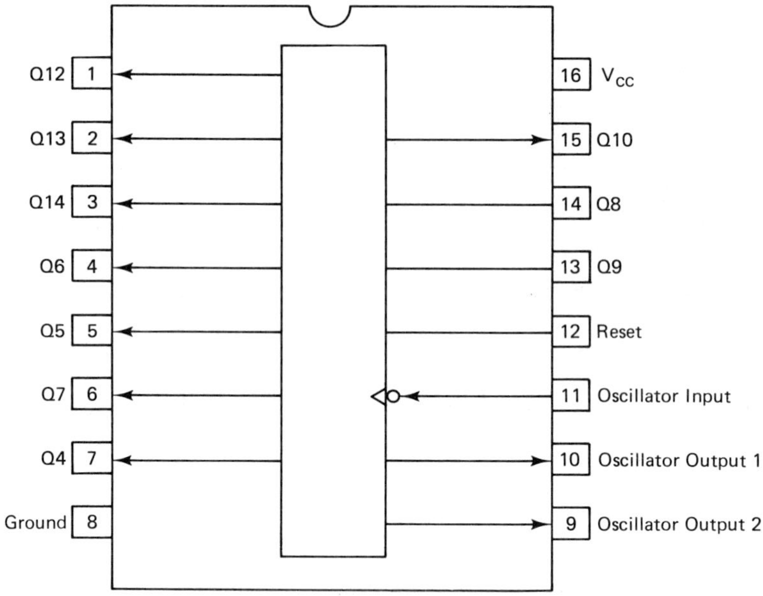

This device is the high-speed CMOS functional equivalent of the 4060 standard CMOS device. It consists of 14 master/slave flip-flops and an oscillator whose frequency is controlled by either a crystal or an external resistor-capacitor circuit. The outputs of the flip-flops are labeled Q4 through Q10 and Q12, Q13, and Q14. The counter advances on the negative edge of the oscillator/clock signal (OCS) applied to pin 11. If the internal oscillator's output is not applied to pin 11, an external clock source may be used. The output of each flip-flop feeds the next, and the frequency at each output is half that of the preceding one. Because of external ripple delays, the outputs do not change simultaneously. There is a reset input at pin 12 that should be kept low for normal operation; if it is high, all outputs go to low. Maximum clock frequency is 5 MHz at 2 V and 25 MHz at 5 V. The function table for this device is:

OCS	RESET	OUTPUT
R	L	No change
F	L	Advance one count
X	H	All low

116 74HC4066 Quad Analog Switch/Multiplexer/Demultiplexer

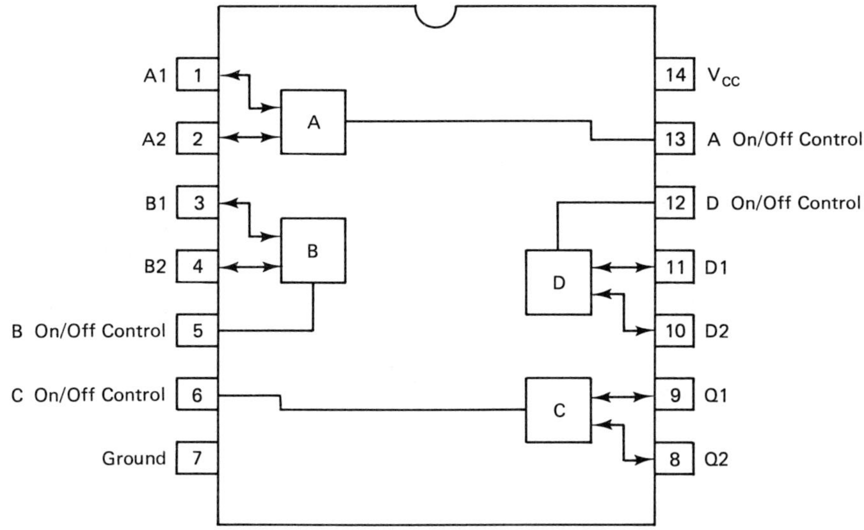

This device is the high-speed CMOS functional equivalent of the 4077 standard CMOS device. It consists of four separate analog switches with a separate on/off control input for each switch. Each switch has two input/outputs. Vcc for this device can range from 2 to 12 V, and input voltages to each switch can range from ground to the Vcc used. This device is identical in function and pin connections to the 74HC4016, but it has lower "crosstalk" between switches and improved linearity across its input voltage range. The function table for this switch is:

ON/OFF SIGNAL	SWITCH
L	Off
H	On

117 74HC4075 Triple 3-Input OR Gate

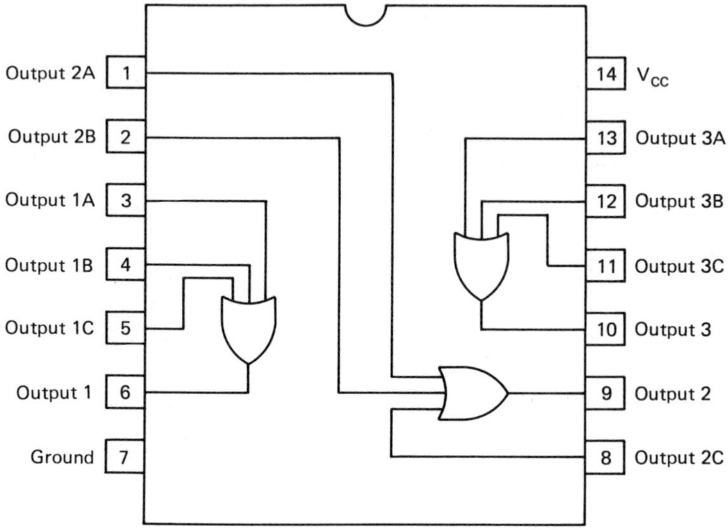

This device is the high-speed CMOS functional equivalent of the 4075 standard CMOS device. It consists of three separate OR gates, each having three inputs. Maximum propagation delay is 115 ns at 2 V and 23 ns at 5 V. Maximum quiescent mode current consumption is 2 μA per gate. The logic table is:

INPUTS			OUTPUT
A	B	C	
L	L	L	L
H	X	X	H
X	H	X	H
X	X	H	H

118 74HC4078 8-Input NOR/OR Gate

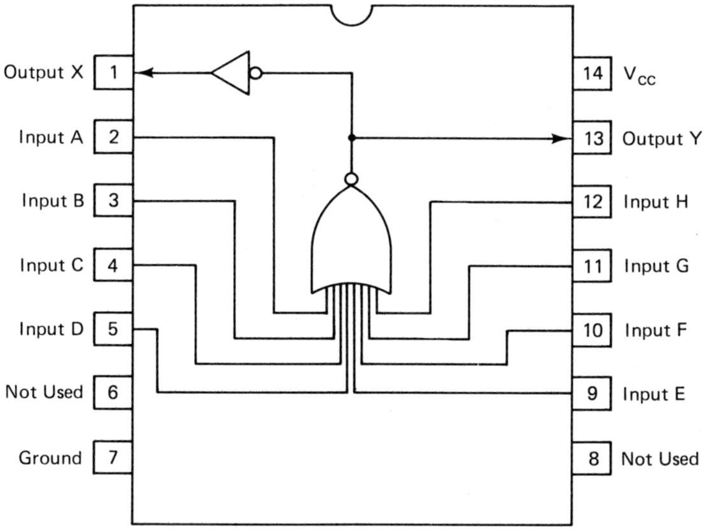

This device is the high-speed CMOS functional equivalent of the 4078 standard CMOS device. There are eight inputs labeled A through H, and two outputs labeled X and Y. Maximum propagation delay is 130 ns at 2 V and 26 ns at 5 V. Maximum quiescent mode power consumption is 2 μA. The logic table is:

INPUTS	X	Y
All low	L	H
Any other combination	H	L

119 74HC4511 BCD to Seven-Segment Latch/Decoder/Display Driver

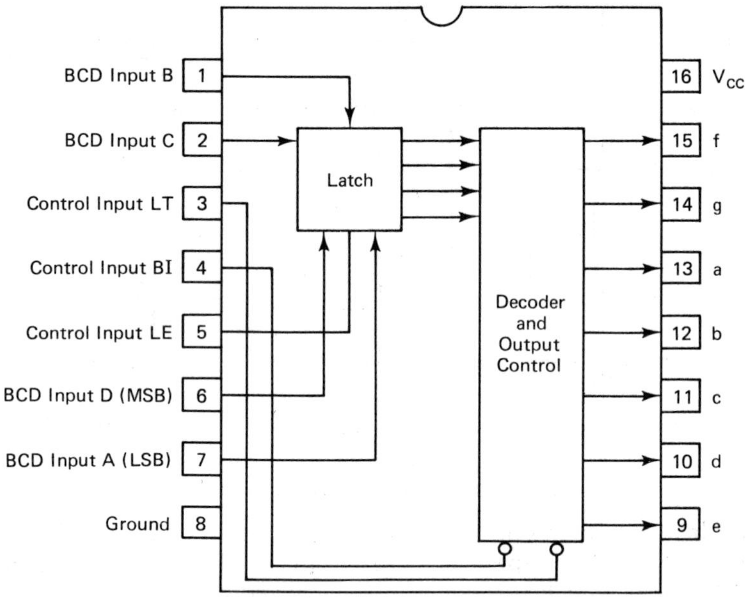

This device is the high-speed CMOS functional equivalent of the 4511 standard CMOS device. It accepts four BCD inputs (labeled A through D) and produces seven segment outputs (labeled a through g). These segment outputs can be used directly with a seven-segment LED numeric display or indirectly with incandescent, fluorescent, gas discharge, or liquid crystal display systems (interfacing circuitry will be required for indirect applications). Three control inputs—lamp test (LT), blanking (BL), and latch enable (LE)—are used to test the display, turn off or pulse-modulate the display brightness, and store a BCD input code. The following table shows the segment outputs lit and the numeral formed when this device is used to drive a seven-segment LED display device:

INPUTS							OUTPUTS							
LE	BL	LT	D	C	B	A	a	b	c	d	e	f	g	Display
X	X	L	X	X	X	X	H	H	H	H	H	H	H	8
X	L	H	X	X	X	X	L	L	L	L	L	L	L	Blank
L	H	H	L	L	L	L	H	H	H	H	H	H	L	0
L	H	H	L	L	L	H	L	H	H	L	L	L	L	1
L	H	H	L	L	H	L	H	H	L	H	H	L	H	2
L	H	H	L	L	H	H	H	H	H	H	L	L	H	3
L	H	H	L	H	L	L	L	H	H	L	L	H	H	4
L	H	H	L	H	L	H	H	L	H	H	L	H	H	5
L	H	H	L	H	H	L	L	L	H	H	H	H	H	6
L	H	H	L	H	H	H	H	H	H	L	L	L	L	7
L	H	H	H	L	L	L	H	H	H	H	H	H	H	8
L	H	H	H	L	L	H	H	H	H	L	L	H	H	9
H	H	H	X	X	X	X	No change from previous state							

120 74HC4514 1 of 16 Decoder/Demultiplexer with Address Latch

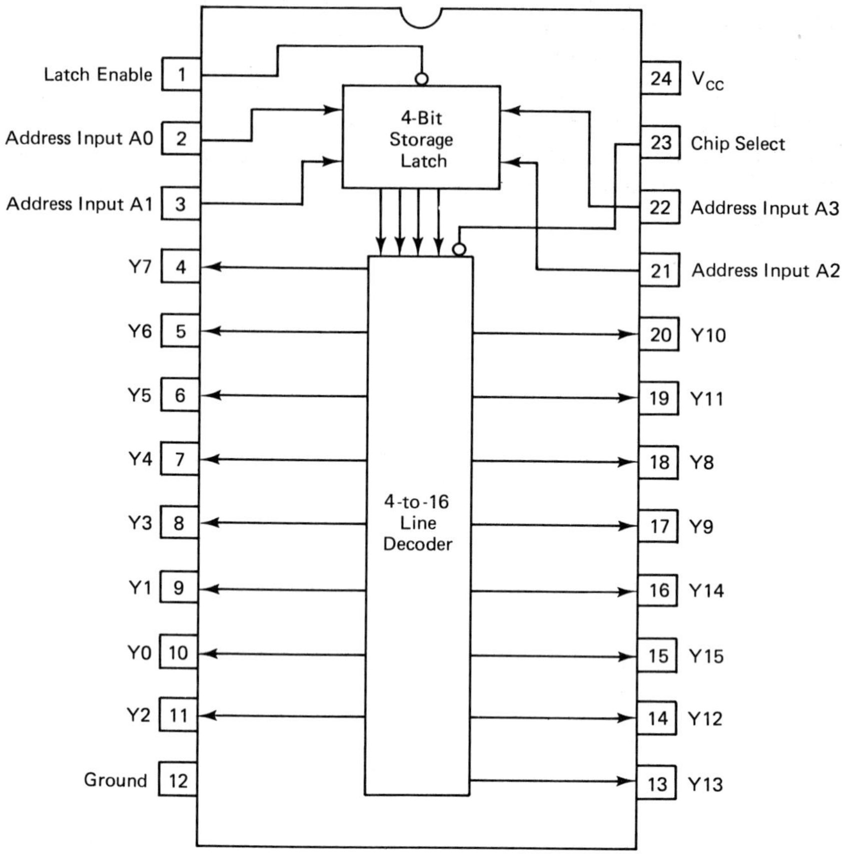

This device is the high-speed CMOS functional equivalent of the 4514 standard CMOS device. It consists of a four-bit storage latch with latch enable (LE) and chip select (CS) inputs and a 4-to-16 line decoder. There are four binary address inputs labeled A0 through A3 and 16 outputs labeled Y0 through Y15. The address inputs are fed to the latch, and when the latch is enabled, are used by the decoder to select one of the outputs and make it high; the unselected outputs remain low. The latch enable input must remain high so that the outputs can follow changes in the address inputs. If it is low, the address inputs are stored in the latch. The chip select input must remain low for normal operation. If it is high, all outputs are set to low. Maximum propagation delay from address inputs to outputs is 230 ns at 2 V and 46 ns at 5 V. Maximum quiescent mode current consumption is 8 μA. The function table for this device is:

LE	CS	A3	A2	A1	A0	OUTPUT
H	L	L	L	L	L	Y0
H	L	L	L	L	H	Y1
H	L	L	L	H	L	Y2
H	L	L	L	H	H	Y3
H	L	L	H	L	L	Y4
H	L	L	H	L	H	Y5
H	L	L	H	H	L	Y6
H	L	L	H	H	H	Y7
H	L	H	L	L	L	Y8
H	L	H	L	L	H	Y9
H	L	H	L	H	L	Y10
H	L	H	L	H	H	Y11
H	L	H	H	L	L	Y12
H	L	H	H	L	H	Y13
H	L	H	H	H	L	Y14
H	L	H	H	H	H	Y15
X	H	X	X	X	X	All outputs low
L	L	X	X	X	X	Latched data

121 74HC4538 Dual-Precision Retriggerable, Resettable Monostable Multivibrator

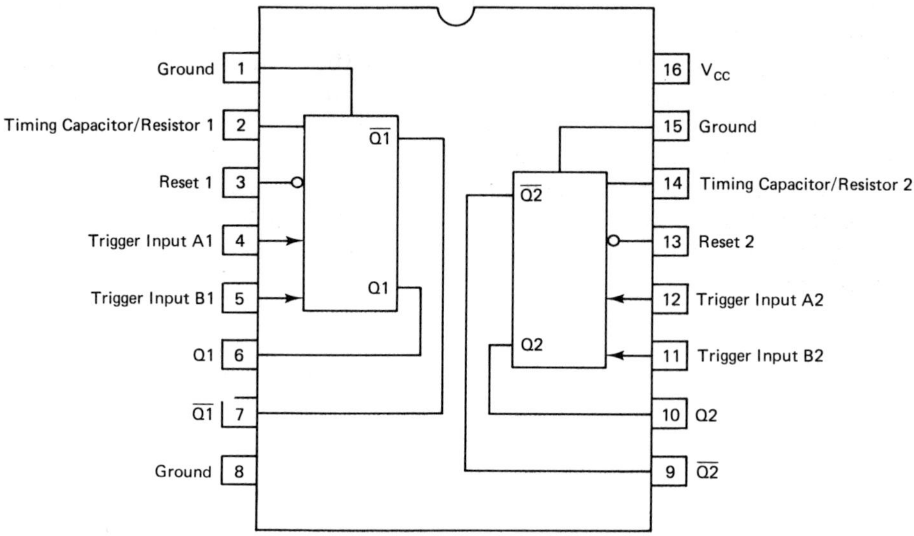

This device is the high-speed CMOS functional equivalent of the 4538 standard CMOS device. It consists of two monostable multivibrators that may be triggered by either the positive or the negative edge of trigger input pulses (labeled A and B) to produce an output pulse over a wide range of widths as determined by external timing resistors and capacitors. Output pulse times can range from 10 μs to ten seconds, although reliable operation drops off if output pulse time exceeds one second. Output pulse width is independent of trigger input pulse width. Each multivibrator also has a reset input that should be held high for normal operation. If it is low, Q will be low and $\bar{Q}$ will be high regardless of the trigger inputs. Using identical external timing components with each multivibrator will results in their output pulses being to within 5% of each other. In the following table, "P" represents a positive output pulse while "N" denotes a negative output pulse. "NP" indicates any input that is not the positive edge of an input pulse.

INPUTS			OUTPUTS	
Reset	A	B	Q	$\bar{Q}$
H	R	H	P	N
H	L	F	P	N
H	X	L	Not triggered	
H	H	X	Not triggered	
H	NP	H	Not triggered	
H	L	NP	Not triggered	
L	X	S	L	H
R	X	X	Not triggered	
F	X	X	Not triggered	

122 74HC4543 BCD to Seven-Segment Latch/Decoder/Display Driver for LCDs

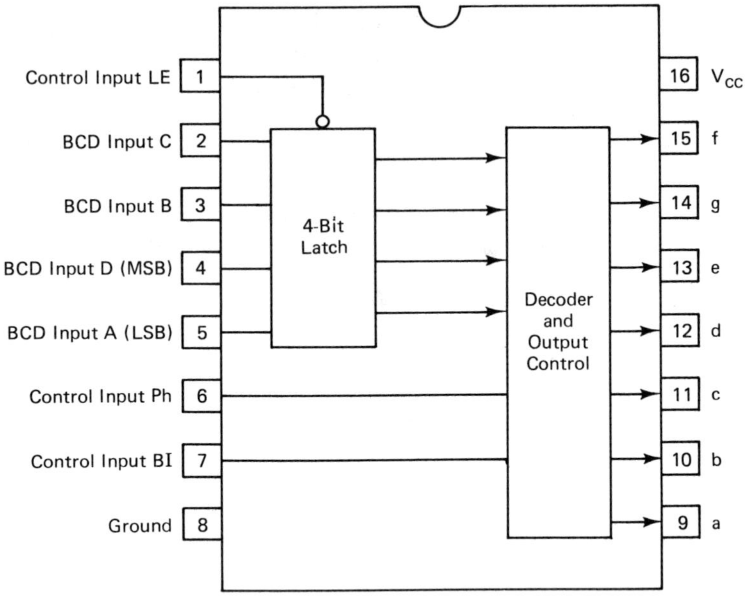

This device is the high-speed CMOS functional equivalent of the 4543 standard CMOS device. It is similar to the 74HC4511 but is intended to directly drive seven-segment liquid crystal display (LCD) units. There are four BCD inputs, labeled A through D, which feed a four-bit latch. The output of the latch goes to a decoder and output control section, which directly drives the LCD segments through seven outputs labeled a through g. The operation of the latch and decoder sections is controlled through three control inputs: latch enable (LE), blanking (BL), and phase (PH). For normal operation, the latch enable should be kept high and the blanking and phase inputs kept low. The maximum propagation delay between the BCD inputs and the outputs is 600 ns at 2 V and 120 ns at 5 V. Maximum quiescent mode current consumption is 8 μA. The function table for this device is:

INPUTS							OUTPUTS							
LE	BL	PH	D	C	B	A	a	b	c	d	e	f	g	Display
X	H	L	X	X	X	X	L	L	L	L	L	L	L	Blank
H	L	L	L	L	L	L	H	H	H	H	H	H	L	0
H	L	L	L	L	L	H	L	H	H	L	L	L	L	1
H	L	L	L	L	H	L	H	H	L	H	H	L	H	2
H	L	L	L	L	H	H	H	H	H	H	L	L	H	3
H	L	L	L	H	L	L	L	H	H	L	L	H	H	4
H	L	L	L	H	L	H	H	L	H	H	L	H	H	5
H	L	L	L	H	H	L	H	L	H	H	H	H	H	6
H	L	L	L	H	H	H	H	H	H	L	L	L	L	7
H	L	L	H	L	L	L	H	H	H	H	H	H	H	8
H	L	L	H	L	L	H	H	H	H	H	L	H	H	9
L	L	L	X	X	X	X	No change from previous state							

3

Basic Logic Building Blocks

All the digital logic circuits, including HCMOS ones, can be broken down into a few simple basic logic gates and flip-flop elements. In fact, it is possible to implement any logic circuit using only NAND gates.

If you have worked with other logic families such as TTL and CMOS, you may be able to quickly review this chapter before proceeding with the rest of the book. But you should not skip it entirely. Although the material in this chapter is elementary, it is essential for fully understanding the operation of and logic behind more complex digital logic devices discussed later in this book.

COMBINATIONAL AND SEQUENTIAL LOGIC

Two terms are used to describe the logic circuits constructed from HCMOS (and all other digital logic families). *Combinational* logic circuits have outputs based on their inputs at any given moment regardless of any prior state or condition of the circuit. By contrast, the operation of *sequential* logic circuits depends on prior states or conditions of the circuit. A combinational circuit can be thought of as reacting "instantly" to an input signal, whereas a sequential circuit seems to be operating in increments or steps. (Given the speed with which a digital circuit can operate, however, a sequential circuit can appear to be reacting "instantly" even though it is incremental in nature.) A sequential circuit in operation can be thought of as a line of stacked dominoes in which the first one is tipped over; each change of state is part of a chain in which each gate or element (except the first) is affected by prior devices and in turn affects each subsequent gate or element (except the last). Neither "combinational" nor "sequential" refers to how simple or complex a logic circuit is; both may be uncomplicated or elaborate.

One problem with sequential logic is coordination of the various gates and elements. In complex sequential circuits, the desired result or operation depends on several different blocks operating in sequence. Coordination of these gates and elements depends on a signal known variously as a *clock*, *enable*, or *trigger*. Without this signal, the devices will not emit proper outputs or respond to input signals. An element or circuit whose operation is governed by a clock signal is called *synchronous*. Circuits, devices, or functions that do not require clock signals are said to be *asynchronous*. These devices, as well as sources of clock or trigger signals, will be discussed later in this book.

FUNDAMENTAL TYPES OF LOGIC GATES

The basic types of logic gates are the *AND, NAND, OR, NOR, inverter* (also known as the *NOT* gate), *buffer* (also known as the *YES* gate), *exclusive OR* (XOR), and *exclusive* NOR (XNOR). Figure 3-1 shows the schematic symbols for each of these gates. In each symbol, only the minimum number of inputs (two except for the inverter and buffer gates) are shown. However, gates other than inverters and buffers may have a theoretically infinite number of inputs, and some logic gate ICs (such as the 74HC30 8-input NAND gate) are available containing gates with more than two inputs. Most logic ICs have multiple independent gates on them, with four being a common number for AND, NAND, etc., gates, while six or eight buffers/inverters may be fitted into a single device package.

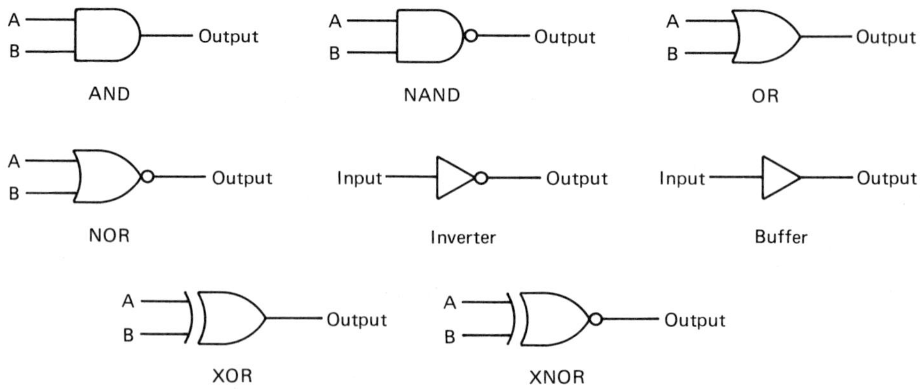

Figure 3-1 Basic Types of Logic Gates

The AND is one whose output is high if *all* inputs (labeled A and B in figure 3-1) are high. If any one, or both, of the inputs are low, the output of an AND gate will be low. By contrast, the functioning of the NAND gate is the opposite of the AND. The output of a NAND gate is high whenever at least one of its inputs is low. If all inputs are high, the output goes low. The logic tables for these gates illustrate this. In these and all subsequent logic tables, "H" represents a high logic level, "L" represents a low level, and "X" represents a condition where the logic state may be high or low.

AND			NAND		
A	B	Output	A	B	Output
H	H	H	H	H	L
X	L	L	H	L	H
L	X	L	L	H	H
			L	L	H

Two other complementary gates are the OR and NOR gates. The OR gate will produce a high output if at least one of its inputs is high, while the NOR gate will produce a high output when *all* of its inputs are low; if one or more inputs are high, the output will be low. The logic tables for these gates are as follows:

OR			NOR		
A	B	Output	A	B	Output
L	L	L	L	L	H
L	H	H	H	X	L
H	L	H	X	H	L
H	H	H			

The inverter and buffer gates are also complementary. The output of the inverter is the opposite of its input; a high input produces a low output, while a low input produces a high output. By contrast, the buffer has the same logic levels at input and output. Inverters are used for logic-level conversions, while buffers are frequently used in interfacing schemes between logic families or to isolate different circuit sections from each other.

The XOR gate produces a high-level output if the number of high-level inputs to it is an *odd* number. For a two-input XOR as shown in figure 3-1, this means the output is high when the inputs are different; they are low when they are the same. The logic table for a two-input XOR gate is:

A	B	OUTPUT
L	L	L
L	H	H
H	L	H
L	L	L

The XNOR gate is functionally equivalent to an XOR gate followed by an inverter. It produces a high output when the inputs are all the same (all high or all low). The logic table for this gate is:

A	B	OUTPUT
L	L	H
L	H	L
H	L	L
H	H	H

All of these gates are available in IC form. In addition, each of these functions may be produced using combinations of other logic gates, NAND gates in particular.

THE 74HC00 QUAD NAND GATE

The 74HC00 device consists of four separate and independent NAND gates in a single package. Any of the logic gates shown in figure 3-1 can be implemented using only NAND gates. Figure 3-2 shows how this is done with some basic gates.

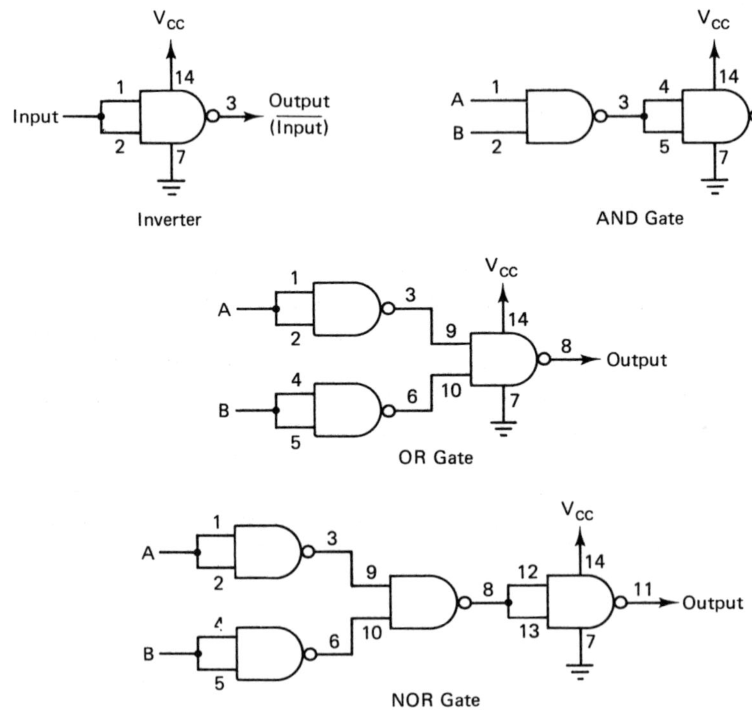

Figure 3-2 Implementing Logic Gates Using the 74HC00

An inverter can be implemented simply by connecting both inputs together and applying the same input signal to both; the output is the opposite of the input. The AND gate is nothing more than an ordinary NAND gate followed by an inverter. As with all AND gates, a high signal is necessary at both inputs to produce a high output. The OR gate uses an inverter at each input; their outputs serve as the inputs for a standard NAND gate. A high signal at either or both inputs will produce a high output signal. Three inverters are used to implement a NOR gate, which is an OR gate followed by an inverter, causing the action of the NOR gate to be the opposite of the OR gate.

Three more-complex logic gates are shown in figure 3-3. The AND–OR gate has four inputs labeled A, B, C, and D. Although each of the four inputs is electrically independent, they are "logically" grouped into two pairs (or sets) consisting of A and B plus C and D. If at least one pair of inputs are both high, the output will be high; otherwise the output will be low. The logic table for an AND–OR gate is:

A	B	C	D	OUTPUT
X	X	H	H	H
H	H	X	X	H
H	H	H	H	H

The XOR gate in figure 3-3 is an interesting design, as each input signal is applied to two different gates and the output of the first NAND gate is used as one input for two NAND gates that use the original input signals as the other input. The XNOR gate requires five NAND gates and thus must be implemented using more than one 74HC00 device. As mentioned earlier, the action of the XNOR gate is the opposite of the XOR gate. Although implementing an XNOR gate using NAND gates is complex, it does illustrate how it is possible to produce any of the basic logic gates using only NAND gates. This is why several 74HC00 devices are useful to have around—in a pinch, they can be substituted for other gates in circuit design and then later replaced by more appropriate devices.

FLIP-FLOP CIRCUITS FROM LOGIC GATES

Flip-flop circuits are among the most useful in digital electronics. These are also known as *bistable multivibrators.* The output of the flip-flop circuit is either a low- or high-level logic signal; there is usually another *complementary* output that has the opposite state of the main output. The main output stays low or high until an input signal known as the *trigger* is received; when it is, the output changes to the opposite state. This is the simplest form of a flip-flop circuit, and most "real-world" flip-flops are more complex in their operation.

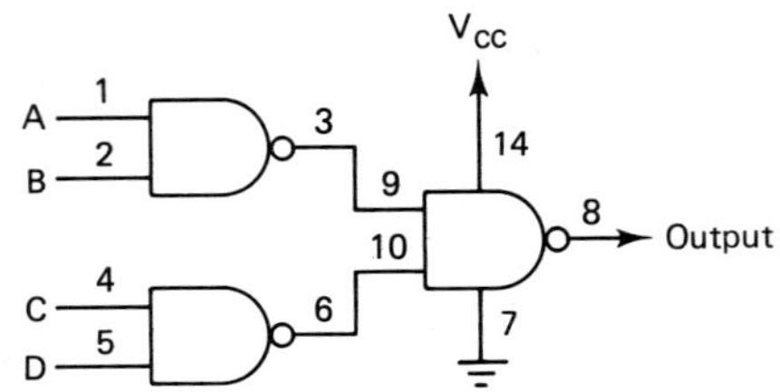

AND-OR Gate

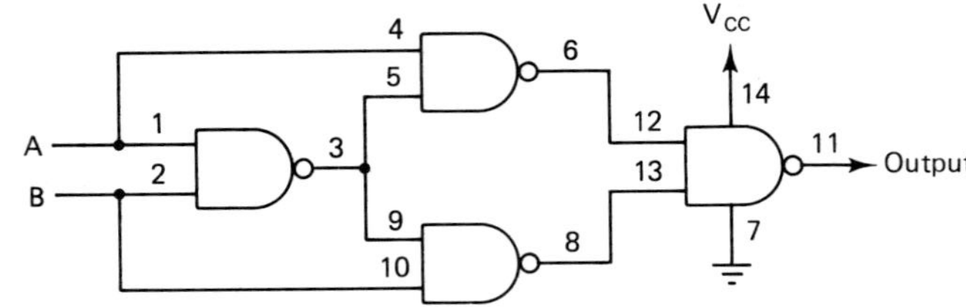

Exclusive OR Gate

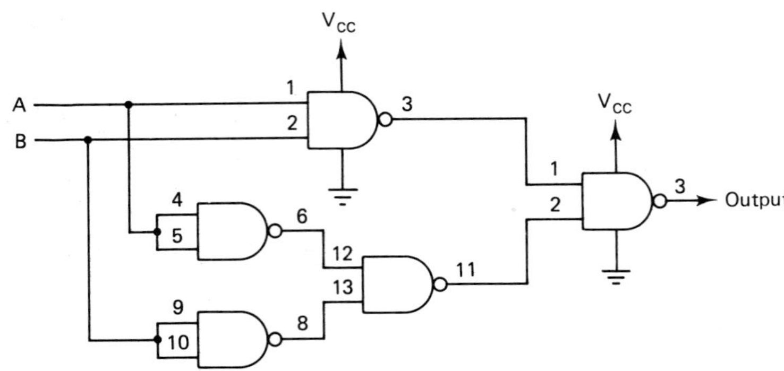

Exclusive NOR Gate

Figure 3-3 Additional Logic Gates Using the 74HC00

ICs containing one or more flip-flop circuits are available, and flip-flop sections are also incorporated into more elaborate integrated devices. These will be examined later in this book. For the moment, we will look at simple flip-flop circuits implemented using NAND gates contained in a 74HC00 IC in order to understand principles applicable to all flip-flop circuits.

Figure 3-4 shows one type of flip-flop known as an *R-S (reset-set)* flip-flop. This circuit is also known as a *latch*. The circuit has an output, Q, and its complement, $\bar{Q}$. Both Q and $\bar{Q}$ can be either high or low, but they must be opposite in state to each other. The states of Q and $\bar{Q}$ depend on the signals preset at the *set* input (S) and the *reset* (R). The operation of this circuit can be illustrated by the following logic table:

S	R	Q	$\bar{Q}$
L	L	Not allowed	
L	H	H	L
H	L	L	H
H	H	No change from previous state	

In the table above, the term "not allowed" refers to a condition where Q and $\bar{Q}$ would have the same logic level, which is not permitted by the definition of a flip-flop circuit. When the outputs of a flip-flop circuit change (switch "states") in response to changes in its input signals, the circuit is said to *toggle*.

The action of the R-S flip-flop shown in figure 3-4 is entirely controlled by the signals at R and S. In many situations, however, it is preferable to have the flip-flop's response to the signals at R and S depend on an enable or clock signal. Figure 3-5 shows an *enabled* or *clocked* R-S flip-flop using four NAND gates. The circuit "ignores" the signals at R and S until a positive pulse or

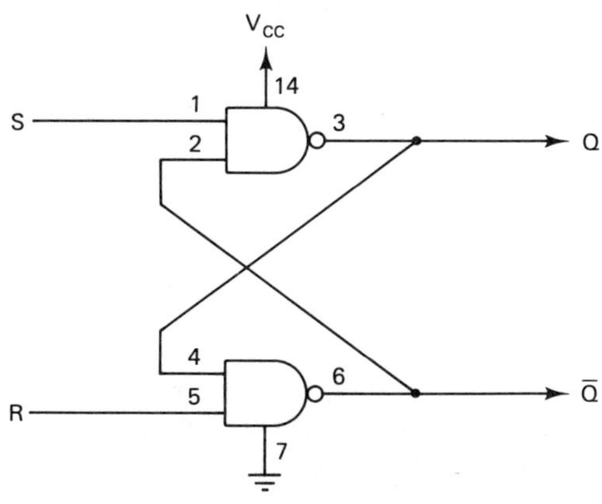

Figure 3-4 R-S Flip-Flop Using the 74HC00

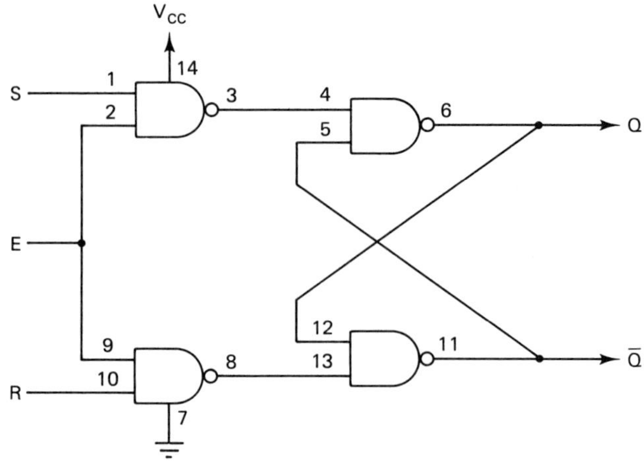

Figure 3-5 Enabled R-S Flip-Flop Using the 74HC00

signal is received at input E. The operation of this circuit is described by the following table:

S	R	Q	$\bar{Q}$
L	L	No change from previous state	
L	H	L	H
H	L	H	L
H	H	Not allowed	

Note that this table is valid *only* when a positive pulse or signal is present at E; otherwise the circuit maintains its previous state.

There is another type of enabled flip-flop that has only a single input, the *D* (for "data") flip-flop. When the E (enable) input is high, output Q will have the same logic level as the signal applied to input D, and $\bar{Q}$ will have the opposite (complementary) logic level. When input E is low, there is no change in Q and $\bar{Q}$. Figure 3-6 shows a D flip-flop built from five NAND gates. The function of this circuit is shown by the following table:

D	E	Q	$\bar{Q}$
X	L	No change	
L	H	L	H
H	H	H	L

The R-S flip-flop in figure 3-4 does not require a clock or trigger signal for operation and is therefore asynchronous. The enabled R-S flip-flop and D flip-flop do and are synchronous circuits.

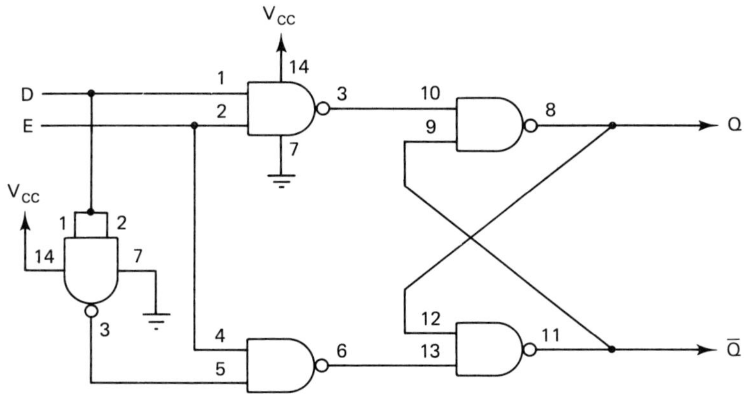

Figure 3-6 D Flip-Flop Using the 74HC00

Flip-flop circuits are actually the most basic examples of memory circuits. The outputs, Q and $\bar{Q}$, can represent one bit of data in binary form; since the output will not change unless new "data" are received in the form of changing input signals, the flip-flop can be said to actually "store" data. As you might expect, the enabled or clocked versions of flip-flops are preferred, since this gives greater control over the "reading" and "storage" of data. Flip-flop circuits may seem crude compared with high-capacity semiconductor memories, but their functioning is remarkably similar. In fact, the earliest computers used flip-flops constructed from vacuum tubes to store and manipulate data.

USING OTHER LOGIC GATE DEVICES

In the previous section, NAND gates were configured so as to provide other logic gate functions, such as NOR and AND gates. Since ICs containing logic gates other than NAND are available, these should be used rather than implementing all logic functions using NAND gates.

Figure 3-7 shows two applications using the 74HC08, which contains four independent two-input AND gates. A buffer is easier to implement, since only one gate is required instead of two. Another application of the 74HC08 is a *transmission* or *enabled* AND gate. This gate has one input connected to a switch that allows the input to be set to high or low. The logic table for this circuit is:

INPUT	SWITCH	OUTPUT
L	L	L
L	H	L
H	L	L
H	H	H

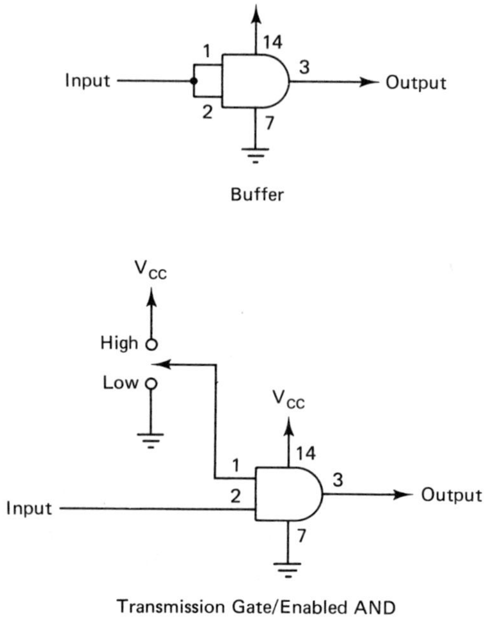

Figure 3-7 Applications of the 74HC08

The 74HC02 device contains four independent two-input NOR gates in the same package. One useful application for this device is shown in figure 3-8. This *pulse stretcher* will produce an output pulse that is a "lengthened" replica of an input pulse applied to the circuit. The length of the output pulse depends on the values of capacitor C and resistor R. The formula for determining the length of the output pulse is 0.8RC, with the value of R measured in ohms and the value of C measured in farads (or, more commonly, microfarads or picofarads). This circuit is also known as a *monostable,* or *"one shot," multivibrator.*

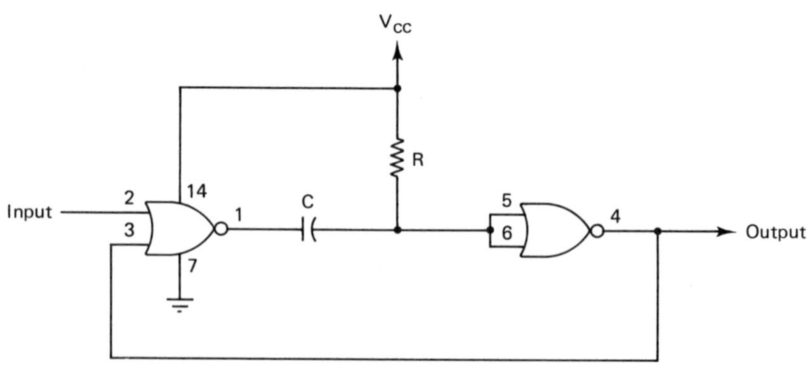

Figure 3-8 Pulse Stretcher Using the 74HC02

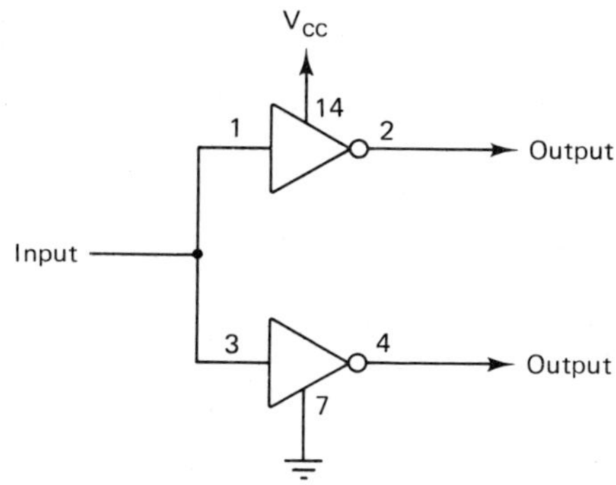

Figure 3-9 Input Expander Using the 74HC04

A widely used HCMOS device is the 74HC04 *hex inverter,* which contains six independent inverter gates. These can be configured as buffers by connecting two inverters in series. Another application is in *input expanders,* as shown in figure 3-9. This takes a single input and produces two outputs that are identical to the input except that the logic level is reversed (inverted). If a noninverted output is desired, either or both outputs may be followed by another inverter. Another application is shown in figure 3-10. This circuit allows selecting a stable, constant high or low output as determined by the switch. It can be particularly useful when testing or experimenting with HCMOS circuits, since constant logic-level inputs are essential when doing so. If you do much experimenting with HCMOS, constructing a few of these circuits will save you much time and aggravation.

APPLICATIONS OF LOGIC GATES

Virtually all the complex devices covered in chapter 2 use logic gate networks to perform their functions. In this section, we will "preview" some of these functions through their logic gate IC equivalents.

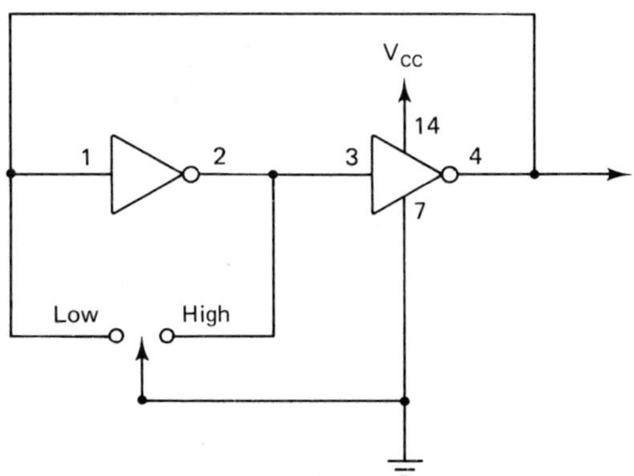

Figure 3-10 Selectable Steady-State Output Circuit Using the 74HC04

The circuit in figure 3-11 is a *half adder* and is the simplest example of a circuit that can take two inputs and produce their sum. All NAND gates are from a 74HC00 device. There are two inputs, labeled A and B, and two outputs known as *sum* and *carry*. If both A and B are low (or 0), their "sum" is also low (0). The sum and carry outputs will thus both be low. If one input (A or B) is high (or 1) while the other is low, their sum is high (1) and the sum output will be high while the carry output will be low. If both A and B are high, this presents a problem. In Boolean arithmetic, the sum of 1 and 1 will be 10, and there is no way to represent this with the single sum output shown in figure 3-11. The solution is provided by the carry output. When both A and B are high, the sum output will be low while the carry output will be high. The high signal at the carry output represents the "1" in the sum (10); the carry output can be used as an input for subsequent adder stages. Most ICs that perform counting, addition, and similar functions have some sort of carry output to allow several devices to be connected (or "cascaded") together to handle numbers or sums larger than those a single device can handle.

The *binary coded decimal* (BCD) code expresses each decimal digit as a four-bit binary number. For example, the decimal number 2 is expressed in BCD as 0010, while decimal 5 is BCD 0101. Figure 3-12 shows a circuit that encodes a decimal input into its BCD equivalent. It uses one 74HC30 eight-input NAND gate, a 74HC20 dual four-input NAND gate, and one section from a 74HC00. The decimal inputs are labeled 0 through 9, and the BCD outputs are labeled A to B. In use, all decimal inputs should be low *except* for the desired decimal input, which should be high. For example, if decimal input 1 is high and the rest low, the corresponding BCD output will have D high and all the rest low. Note that decimal input 0 is an open circuit, since no decimal input at all will produce an "all low" BCD output.

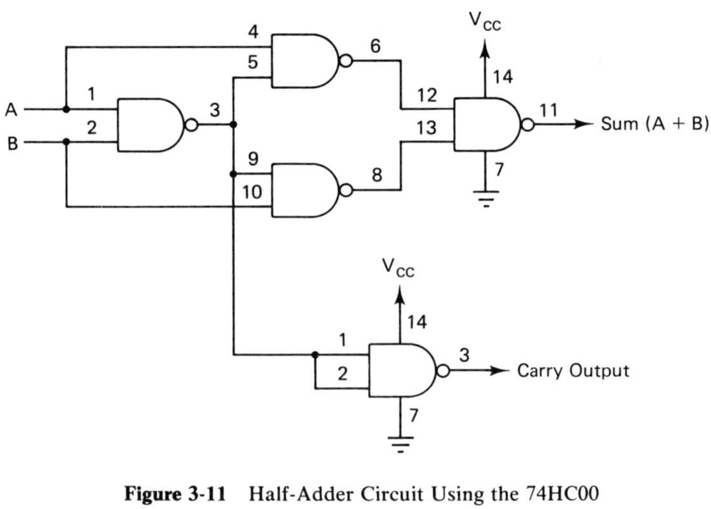

Figure 3-11 Half-Adder Circuit Using the 74HC00

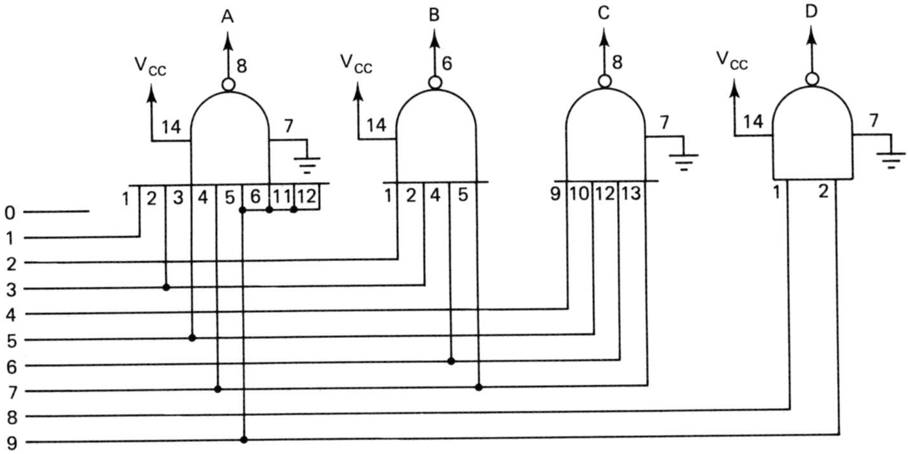

Figure 3-12 BCD Encoder Circuit Using NAND Gates

Decoding BCD inputs and producing a binary or decimal output from them is a bit more complex. Figure 3-13 shows a simple BCD decoder that can recognize a BCD input of 0000. If inputs A through D are all low, the output of the circuit will be high; any other combination will produce a low output. The NAND gate used is one section of a 74HC20 dual four-input device.

A *multiplexer,* or *data distributor,* is a device that takes an input signal and routes it to one of two or more possible outputs. The output is selected by

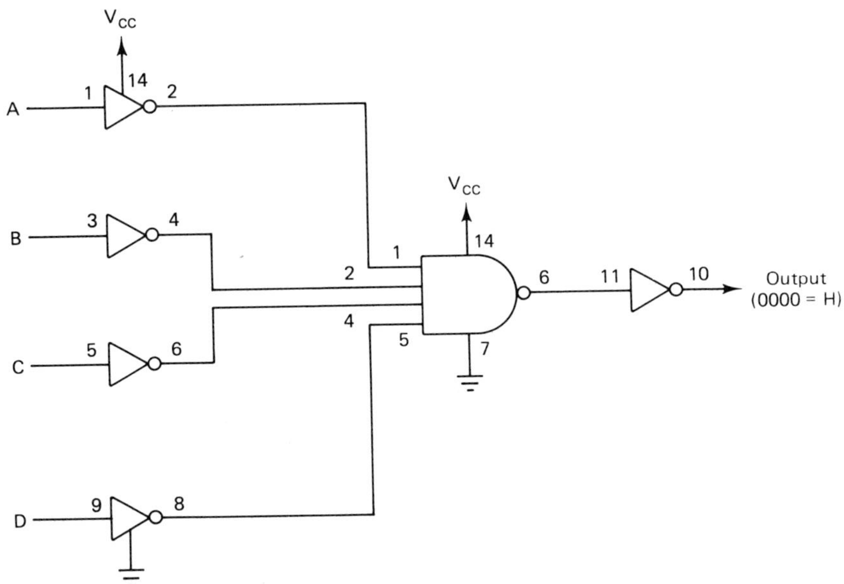

Figure 3-13 BCD Decoder Circuit

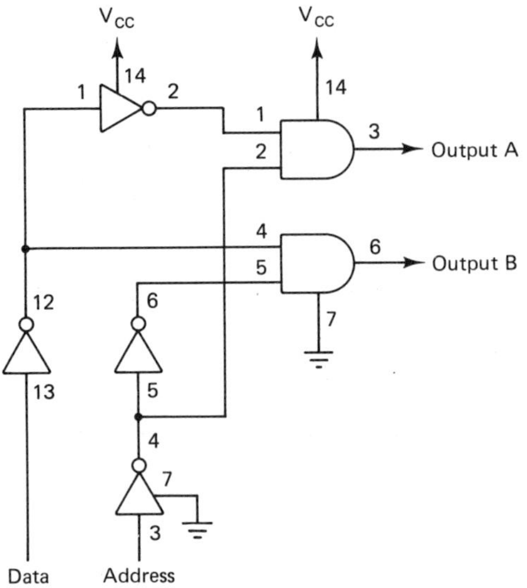

Figure 3-14 Simple 1 of 2 Multiplexer Circuit

one or more *address* inputs. A simple multiplexer is shown in figure 3-14, which uses a 74HC04 hex inverter and a 74HC08 squad AND gate. The function table for this circuit is:

DATA	ADDRESS	A	B
L	L	L	L
H	L	H	L
L	H	L	H
H	H	L	L

The opposite of a multiplexer is a *demultiplexer,* also known as a *data selector.* This circuit selects one of two or more input signals and routes it to a single output. As with multiplexers, the routing is controlled by an address input. Figure 3-15 shows a simple demultiplexer that uses 74HC04 inverters, 74HC08 AND gates, and a 74HC32 OR gate. The function table for this circuit is:

ADDRESS	A	B	OUTPUT
L	X	L	L
L	X	H	H
H	L	X	L
H	H	X	H

Multiplexers and demultiplexers can be designed to handle more than one bit at a time. Figure 3-16 shows a demultiplexer that can choose between two four-bit (nibble) inputs (labeled 1A through 4A and 1B through 4B) and send the appropriate input to the outputs labeled 1 through 4. This circuit

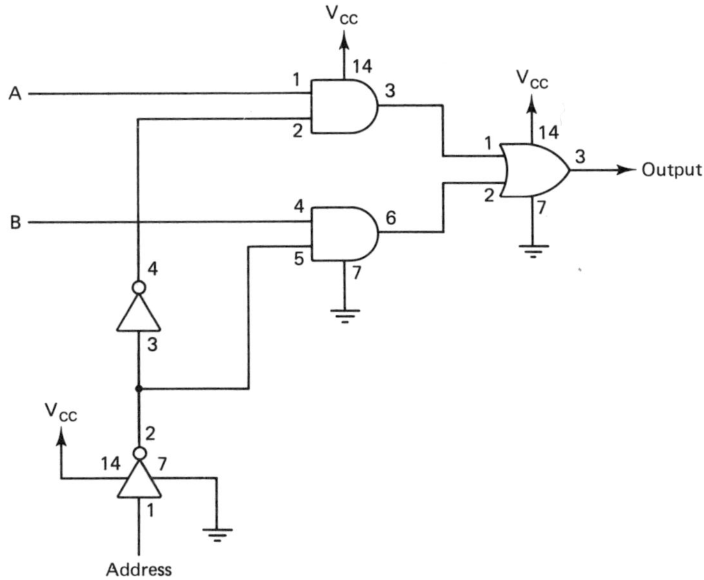

Figure 3-15 Dual-Input Demultiplexer Circuit

uses two 74HC51 dual AND–OR gate ICs and a single inverter from a 74HC04 device. The input selected appears at the output in inverted form. If the address input is low, inputs 1A through 4A are selected, while inputs 1B through 4B are selected if the address input is high.

Multiplexer and demultiplexer ICs are available and were described in chapter 2. Their internal circuitry is remarkably similar to that shown in figures 3-14 and 3-15, however.

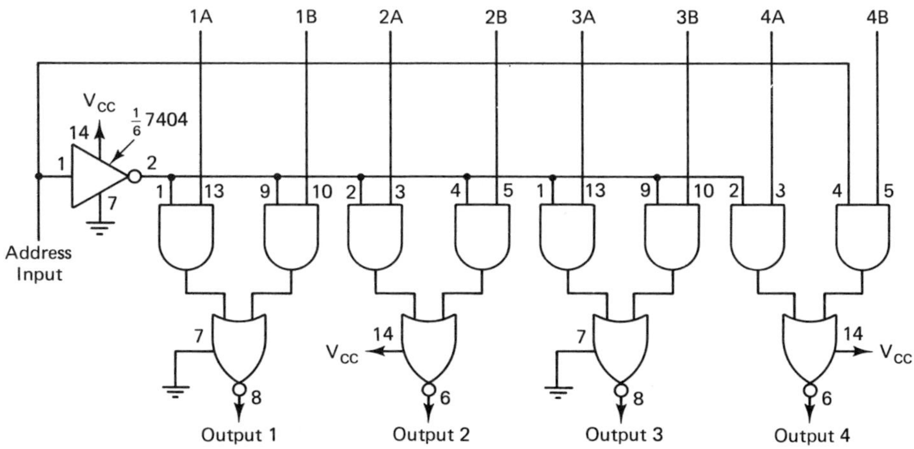

Figure 3-16 1 of 2 Data Selector Using the 74HC51

4

Applications of the 74HCU04 Hex Unbuffered Inverter

The 74HCU04 hex unbuffered inverter device is, at the time this book is being written, unique among HCMOS devices, since it is the only one identified by "HCU." This indicates an unbuffered device with only one stage between input and output. Unlike most other HCMOS devices, it is ideally suited for such linear applications as oscillators, pulse shapers, and high-input impedance amplifiers. It also makes an excellent high-speed digital switch. In this chapter, we will look at some of the applications of this versatile device.

OSCILLATOR APPLICATIONS

Figure 4-1 shows the basic configuration for an oscillator using two sections of the 74HCU04. The output of this circuit is a series of discrete pulses formed by switching the output level from low to high and back again. This type of circuit is known as an *astable* (or *free running*) *multivibrator*. The operating frequency depends on the values of the resistor labeled R and the capacitor C, according to the formula

$$2.2 * R * C$$

The minimum value of R should be 100 ohms, and the minimum value of C should be 100 pF. Moreover, if C is less than 1000 pF, stray capacitance in the circuit may adversely affect operation. The value of R should not be too great in relation to C. The resistor labeled Rs is used to ensure that the output

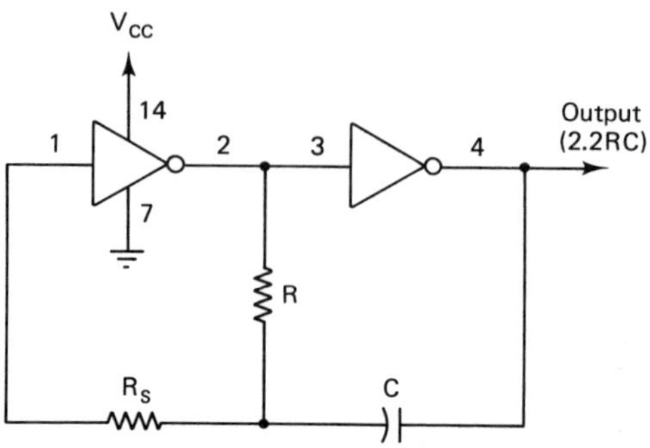

Figure 4-1 Basic Oscillator Configuration for the 74HCU04

frequency will be independent of power supply variations. In many cases, Rs will not be necessary. If Rs is used, it should be of a small value, and smaller in value than R in any case. The output frequency of this circuit will be relatively insensitive to changes in temperature.

The precise output frequency will vary with the tolerance of the component values and the layout and construction of the circuit itself. However, the following are typical output frequencies for the component values specified:

VALUE OF R	VALUE OF C	FREQUENCY
470 ohms	100 pF	6 MHz
4.7 K	100 pF	715 kHz
47 K	1000 pF	8.6 kHz
22.1 K	2.2 μF	80 Hz

It is also possible to use a piezoelectric crystal to control the output frequency of an oscillator circuit. Figure 4-2 shows a circuit with an output frequency of 3.58 MHz as determined by a crystal "cut" for that output frequency. The crystal is a common "color burst" type. The output frequency

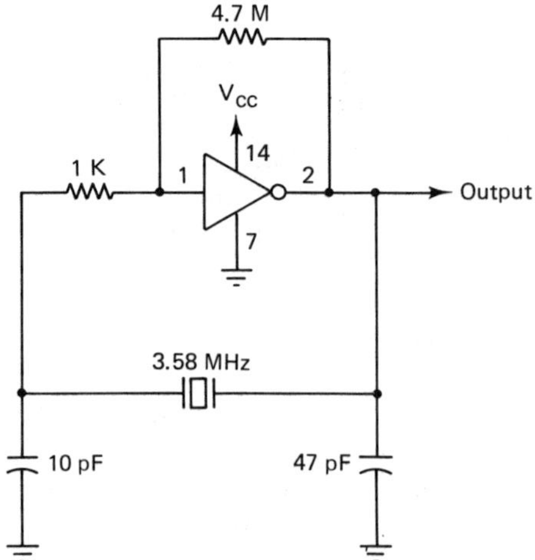

Figure 4-2 Crystal-Controlled 3.58 MHz Oscillator Using the 74HCU04

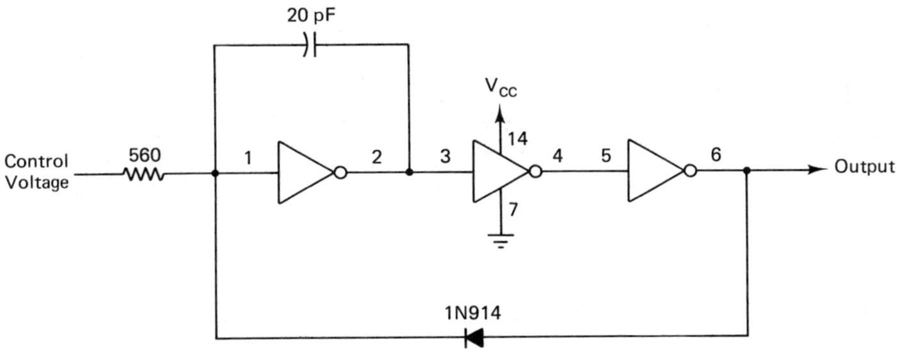

Figure 4-3 73HCU04 High-Frequency VCO Pulse Generator

of this circuit is highly stable, making it suitable for use as a clock/trigger signal source or for precision-timing applications. It is possible to vary the output frequency of this circuit somewhat by replacing the 10 pF capacitor with a varactor and applying a DC control voltage to the varactor through a current limiting resistor. This circuit is designed to operate from a Vcc of +5 V, and the output will be a series of square wave pulses alternating between Vcc and ground.

The 74HCU04 can also be used in a *voltage controlled oscillator* (VCO). As the name indicates, this is a type of oscillator whose output frequency depends on the value of a control voltage applied to it. Figure 4-3 shows a VCO whose output frequency can be varied from 5 to 25 MHz by varying the control voltage from +2.5 to 6 V. Note in the schematic how some of the oscillator's output is fed back, through a 1N914 diode, into the control voltage input. This helps stabilize the oscillator at higher frequencies and during variations of the supply voltage.

It is also possible to produce oscillators having *negative* pulse outputs (that is, transitions from a high to a low logic level). Figure 4-4 shows a VCO

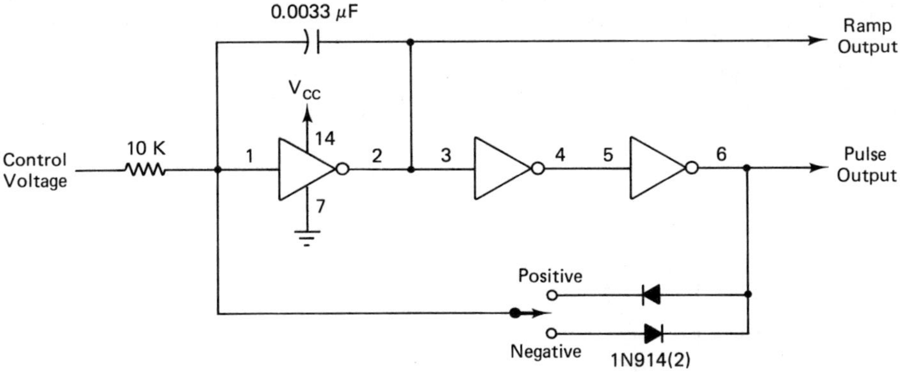

Figure 4-4 Pulse and Ramp Generator Using the 74HCU04

circuit that can produce pulse and ramp outputs in both positive and negative form. This switching between positive and negative outputs is accomplished by selecting, using a switch, between two 1N914 diodes that are opposite in polarity. In the "positive" position, the output frequency can range from 180 kHz to 15 kHz for a control voltage between −2 and +2 V. In the "negative" position, the output frequency can range from 15 kHz to 180 kHz for a control voltage between 2.6 and 7 V.

AMPLIFIER APPLICATIONS

While not intended for use in amplifiers, the 74HCU04 can be used for such applications. Figure 4-5 shows a generalized model of a linear amplifier built around a single inverter. The gain of this circuit depends on the values of resistors R1 and R2, and it is designed to operate into a typical load impedance of 1 K and a load capacitance of 20 pF. The following table illustrates the gain of the circuit, in dB, for different values of R1 and R2:

GAIN	R1	R2
30	0	1 M
21	1 K	1 M
12	1 K	7.5 K
3	250 ohms	1.5 K

The circuit in figure 4-5 may require an input capacitor when used for audio signals.

Figure 4-6 shows an audio or tone amplifier that uses three sections of the device to produce an amplifier capable of driving a 32 ohm speaker or headset. Note that two sections of the 74HCU04 are connected in parallel, and up to five inverter sections may be connected in parallel for additional gain. If five inverters are connected in parallel, the output voltage can reach 3 V peak to peak. The audio quality and bandwidth of this circuit are not great but are suitable for tones or "beeps."

Figure 4-7 shows a unity gain amplifier used as a low-pass filter. The circuit shown reduces all frequencies above 5 kHz by 3 dB through the capaci-

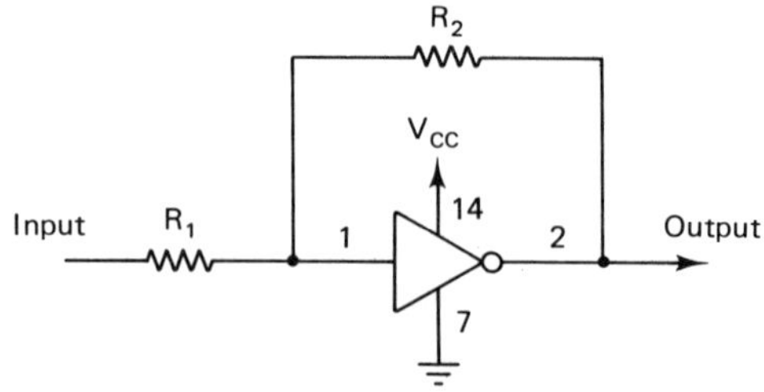

Figure 4-5 Linear Amplifier Using the 74HCU04

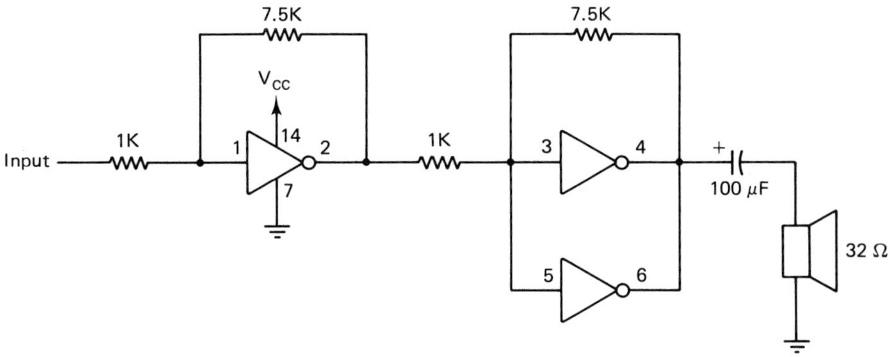

Figure 4-6 Audio/Tone Amplifier

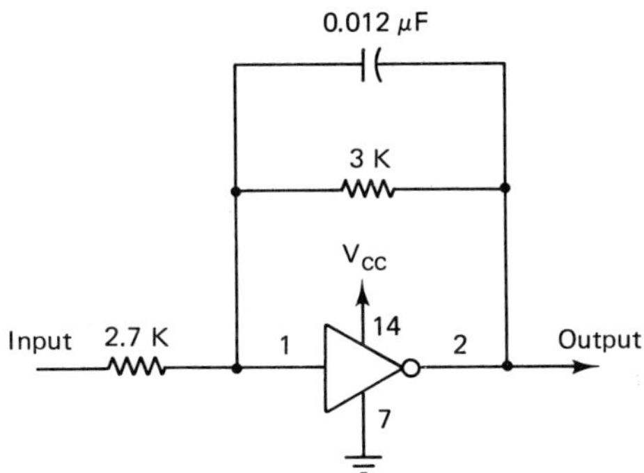

Figure 4-7 Unity Gain Low-Pass Filter

tor feedback arrangement. Additional poles may be added with additional filter stages.

INTERFACING AND DRIVE APPLICATIONS

Unlike other HCMOS devices, the 74HCU04 cannot directly drive TTL devices. Figure 4-8 shows an arrangement by which TTL devices can be directly

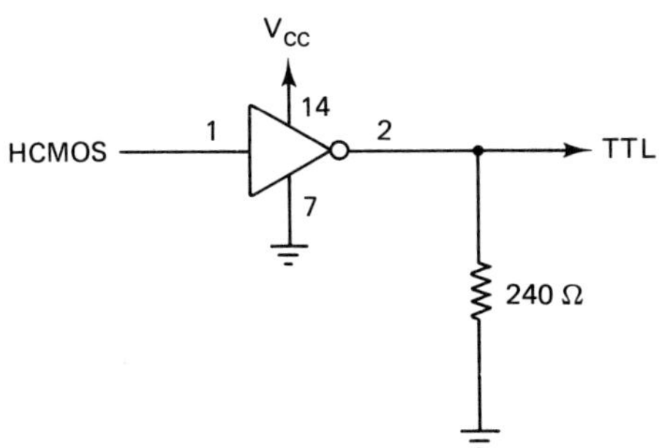

Figure 4-8 Interfacing Between HCMOS and TTL

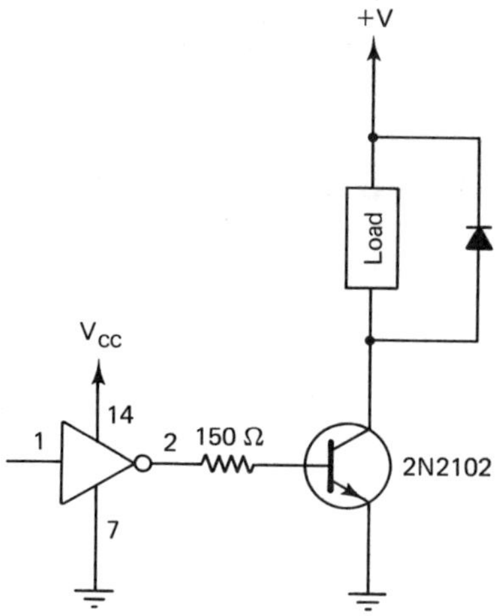

Figure 4-9 Driving External Devices Such As Lamps or Relays

driven from 74HCU04 inverters. Both the TTL device and the 74HCU04 inverter must be operated from a +5 V supply voltage.

A useful application of the 74HCU04 consists of driving external devices such as lamps, relays, solenoids, or similar items that require substantial drive current. This amount of current is beyond the capabilities of the 74HCU04, which can provide a maximum of only 25 mA, so the output of the device is used to control a transistor capable of supplying the additional current. The circuit in figure 4-9 uses a 2N2102 or equivalent transistor to supply 600 mA of current to an external load. The load should have a diode in parallel as shown to prevent damage to the output transistor when the load (the device driven) changes states (such as when a relay opens and closes).

5

Clocked Logic Elements

Perhaps the most important HCMOS devices are those that are *clocked*, meaning that their operation depends on a clock, trigger, or enable signal. Simple circuits of this type built from logic gates were briefly discussed in chapter 3. In this chapter, we will examine more complex circuits and devices whose operation depends on a clock signal.

CLOCKING METHODS

Clock signals are not alike. There are two basic types of clock signals, level and edge clocking. In *level clocking*, the level of the clock signal (whether it is high or low) determines the clocking action. In *edge clocking*, the *transition* between logic levels (from low to high or high to low) determines the clocking action. In addition, both types of clocking may be *positive* or *negative*. For example, in level clocking, either a high or a low logic level may be recognized as a clock pulse. Figure 5-1 shows typical waveforms for these different types of clock signals.

The waveforms shown at (a) and (b) in figure 5-1 are known, for obvious reasons, as *square waves*. The signals switch logic levels abruptly with minimal transition time. The logic levels are represented by voltages, and normal practice is to use the supply voltage for high logic and 0 V (ground) to represent low. Positive level clocking involves switching from 0 V to the supply voltage and back to form a clock pulse, while negative level clocking shifts from the supply voltage to 0 V and back to make a clock pulse.

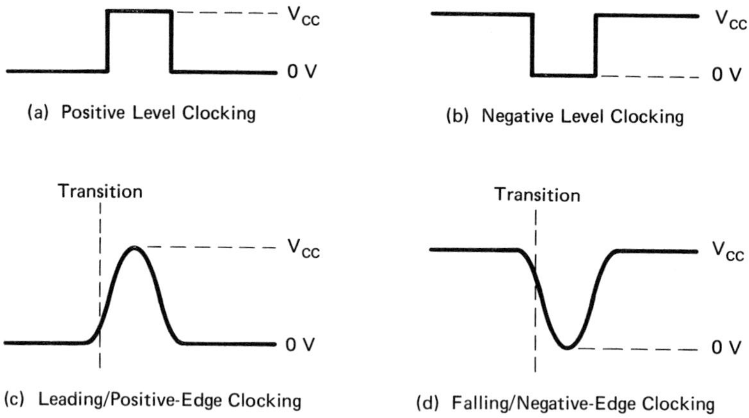

Figure 5-1 Clocking Methods and Signal Levels

The waveforms at (a) and (b) in figure 5-1 are "idealized," since few real-world clock signals even approximate a true square wave. More commonly found, especially at high clock signal frequencies, are the types of waveforms shown at (c) and (d) in figure 5-1. Note that the transition between logic levels is not a "right angle," and the level is not entirely flat. In edge clocking systems, it is the *transition* between logic states that provides the clock signal. The approximate point where the transition occurs is indicated in (c) and (d). Positive-edge clocking is also known as *leading* or *rising* edge clocking and is shown at (c). Negative-edge clocking is also known as *trailing* or *falling* edge clocking and is shown at (d).

If level clocking is used, the clock signal typically must hold at the "triggering" level long enough for the clocking or enabling action to take place. In edge clocking, the duration of the clock pulse is irrelevant; the change between logic levels by itself provides the triggering action.

One major difference between older TTL devices and their HCMOS equivalents involves the clocking methods used. TTL used level clocking for some devices and edge clocking for others, while HCMOS relies (at the time this is being written) exclusively on edge clocking. This applies to HCMOS equivalents of TTL devices that use level clocking. For example, the 7473 is a TTL dual J-K flip-flop device whose outputs can change whenever the clock input signal goes from a low level to a high level. By contrast, the 74HC73 HCMOS equivalent has the same pin connections, internal logic, and functions but uses negative-edge clocking. In actual operation, there is very little real difference between edge or level clocking except that devices using level clocking can generally operate at faster speeds and with less than perfectly formed clock pulses. (On TTL devices that used level clocking, input data could be changed only immediately after clocking took place; input data may be changed anytime if edge clocking is used.) The proper clocking method of each HCMOS device was described in chapter 2.

While HCMOS is much less "fussy" than TTL about what constitutes a clock signal, it is still best to use as "clean" a clock signal as possible. A clean clock signal is one that falls completely to 0 V at a low logic level and rises to V_{cc} at high logic levels. Although HCMOS devices can recognize voltages above 0 V as low level and less than the supply voltage as high level, using 0 V and V_{cc} ensures maximum possible immunity from false triggering. Each clock pulse should have only one peak (either positive or negative) per input cycle; jagged peaks or "ragged" waveforms can be recognized as more than one clock pulse. Decoupling capacitors should also be used to suppress voltage spikes which can be "mistaken" for a clock input signal by a device.

GENERATING CLOCK SIGNALS

Although clock signals may be generated using the circuits described in chapter 4, a simpler and more reliable method (especially at lower frequencies) is to use a circuit built around a CMOS equivalent of the popular 555 timer device. One such IC is the ICM7555 from GE/Intersil. The outputs of this device can drive TTL and CMOS, making it ideal to use with either HC or HCT versions of HCMOS. The supply voltage can range from +2 to 18 V, and it can generate a clock signal up to 1 MHz. The pin connections and internal logic of the ICM7555 are identical to those of a conventional 555 timer.

Figure 5-2 shows the ICM7555 configured to deliver two identical outputs. Both will be positive-edge signals, although an inverter can be added to either or both outputs to provide a negative-edge signal if desired. The formula for determining the output frequency of this circuit is

$$\text{output frequency} = 1/(1.4 * R * C)$$

in which R is the resistor and C is the capacitor shown in Figure 5-2.

The supply voltage for this circuit should be the same as that of the device(s) to which the clock input is provided.

FLIP-FLOP DEVICES

In chapter 3, we discussed elementary flip-flop circuits implemented using logic gates. Flip-flop devices are also available, and flip-flop circuits are used extensively in other HCMOS devices. Indeed, flip-flops may well be the most important category of clocked logic devices. Figure 5-3 shows schematic symbols for the three most important types of flip-flops; you will see these symbols in circuits using integrated flip-flops and in the internal diagrams of integrated flip-flops and other devices using them.

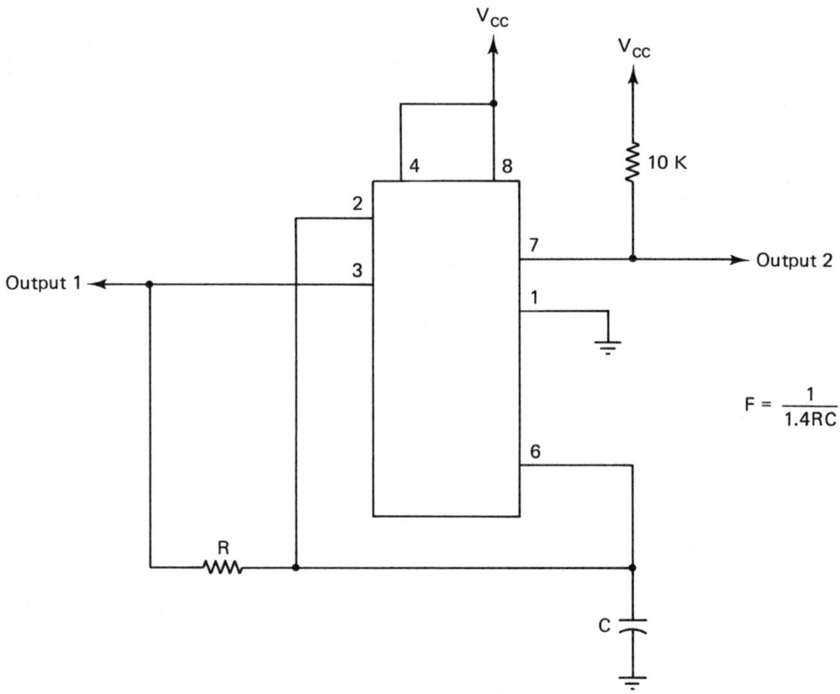

Figure 5-2 Astable Multivibrator/Clock Signal Source Using the ICM7555

The *J-K flip-flop* is similar to the R-S flip-flop examined in chapter 3; the J input corresponds to the S input, and the K input to the R input. What the "J" and "K" actually stand for has been lost somewhere in electronics history, but they do help differentiate this circuit from older R-S flip-flops. Actually, there are several major differences between the two. In function, the J-K flip-flop more closely resembles the enabled R-S flip-flop than the "classic" R-S circuit. The clock input is similar to the enable input. In addition, most (but not all) J-K flip-flops include two new inputs known as *preset* and *clear*. These are included because the outputs, Q and $\bar{Q}$, assume states at random when power is first applied to the circuit. The clear input signal is equivalent to a reset (R) input, while the present input signal is the same as a set (S) input. The preset and clear inputs are used at times other than the initial application of supply voltage; for example, in arithmetic circuits they may be used to enter initial values or clear the count to zero.

The preset and clear inputs are special in that they are *asynchronous*. This means that they affect the circuit regardless of the clock input signal. (The effects of the J and K inputs depend on the clock signal and are therefore *synchronous* inputs.) Thus, attention must be paid to keeping stray or transient signals off the clear and preset inputs.

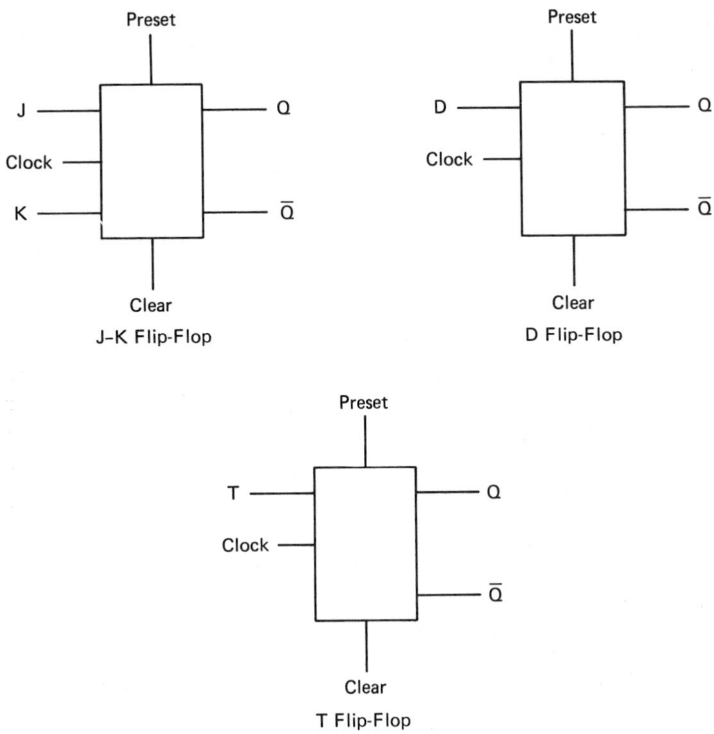

Figure 5-3 Schematic Symbols for Different Flip-Flops

The operation of a typical J-K flip-flop, such as the one shown in figure 5-3, can be illustrated by a logic table. The following table shows the outputs depending on the preset (*pre*), clear (*clr*), clock (*clk*), J, and K signals. "X" is used to indicate that the state of the input—high or low—is irrelevant. "E" indicates that the proper clock signal ("enabling" the circuit) is being received at the clock input.

PRE	CLR	CLK	J	K	Q	$\bar{Q}$
L	H	X	X	X	H	L
H	L	X	X	X	L	H
H	H	E	H	L	H	L
H	H	E	L	H	L	H
H	H	E	H	H	Outputs switch states ("toggle") in response to incoming clock pulses	

From this table, it is apparent that the clear and preset inputs must be held to a high logic level for normal operation of the circuit.

Figure 5-3 also shows the schematic symbol for a D flip-flop. This is very similar to the J-K flip-flop except that there is only one input, D, rather than two (J and K). The clear and preset inputs function as they do for the J-K circuit, and the table of its operation is as follows:

PRE	CLR	CLK	D	Q	$\bar{Q}$
L	H	X	X	H	L
H	L	X	X	L	H
H	H	E	H	H	L
H	H	E	L	L	H

The final schematic symbol in figure 5-3 is for a *T flip-flop* (sometimes known as an *R-S-T flip-flop*). The "T" stands for "toggle," and, as the name implies, the outputs change states in response to the T input. In practice, a T flip-flop is formed by connecting the inputs of a J-K flip-flop. Currently, no T flip-flop HCMOS ICs are available, since they are so easy to implement from J-K flip-flops, but these circuits sometimes show up in the internal diagrams for other ICs. The function of a T flip-flop is simple; each time the T input is high and the circuit is enabled, the outputs toggle.

You will often see flip-flop schematic symbols similar to the J-K flip-flop shown in figure 5-4. At first glance, it may look similar to the J-K flip-flop symbol shown in figure 5-3. However, note that the clock, preset, and clear inputs terminate in small circles when they meet the body of the device. This indicates that these inputs are *negative* or *inverted*. In this case, the signal at the clock input is normally high, and a clock signal is provided when the input drops from high to low (a negative pulse). The inverted clear and preset inputs operate in the opposite fashion from a "normal" flip-flop; a low preset signal sets the circuit, and a low clear signal clears the flip-flop. It is possible for only one input, such as the clear or clock input, to have inverted or negative triggering. Some flip-flop ICs are designed in this way, and negative triggering can easily be implemented by preceding an input with an inverter.

Flip-flops are the essential elements in such circuits as counter/dividers, latches, and shift registers. In this chapter, we will look at how such circuits are implemented using various flip-flops.

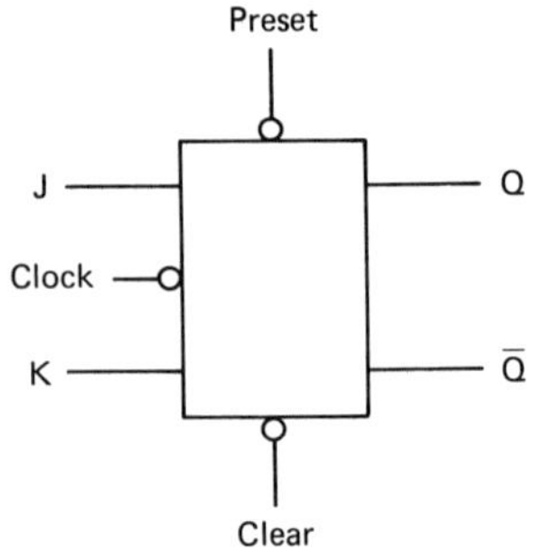

Figure 5-4 Schematic Symbol for J-K Flip-Flop with Inverted Inputs

COUNTER/DIVIDERS AND THEIR APPLICATIONS

One common application of flip-flops is in *counter/dividers.* Figure 5-5 shows a *divide-by-2 counter/divider* which uses a 74HC73 dual J-K flip-flop. This device was described in chapter 2, and we will not repeat that information here. The important thing to note is that both J-K flip-flops in the 74HC73 have a clear input but *do not have a preset input.* This means that the flip-flops can be cleared at any time, but a preset value cannot be entered into either one. The circuit in figure 5-5 shows one of the J-K flip-flops configured so that both the J and K inputs are "tied" to Vcc and held at a high logic level along with the clear input. In this arrangement, the outputs will toggle in response to each clock pulse.

Each clock pulse will cause Q and $\bar{Q}$ to switch states. Thus, the output at Q will be high on every other clock pulse; during the other pulse, $\bar{Q}$ will be high. However, $\bar{Q}$ is not used in this case, meaning that a high output is available (at Q) on only *half* the input pulses. If the clock frequency is 100 Hz, a high output will be available from this circuit at only half that rate, or 50 Hz. In effect, the clock frequency has been "divided" in half. (What really happens, of course, is that the other half of the high-level outputs available at $\bar{Q}$ have been discarded.)

It is possible to view this circuit as an *automatically recycling counter.* It is admittedly a limited counter—it can only count 0 and 1—but it does indeed count the incoming clock pulses in a binary manner. Suppose the output has been set to 0 (output Q at a low level) through the clear input. The first clock pulse will send the output high (the first clock pulse is counted) while the second sends the output low again (the second pulse is counted). The third clock pulse again sends the output high, and the counting is repeated. This shows why we can define *counter* as a circuit that produces one output pulse for a given number of input signals. Since the output is but a fraction of the input signals, we can also say the circuit divides the input signals. This is why such circuits are referred to as counter/dividers.

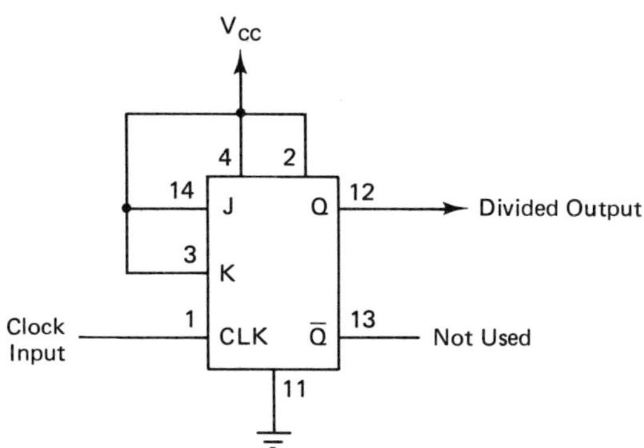

Figure 5-5 Divide-by-Two Counter Using the 74HC73

A circuit such as the one in figure 5-5 is not very useful. However, it is possible to cascade several flip-flops together to count higher. Figure 5-6 shows a divide-by-4 counter/divider using both halves of the 74HC73. Both halves share a common clock signal. The first flip-flop produces an output that is half the clock frequency. This output in turn serves as the J and K inputs for the other flip-flop. The result is that outputs X and Y can together have four different states. Both start at low, and then X goes high while Y remains low. Then they toggle as Y goes high while X goes low. Finally both go high and then "recycle" back to both being low. The clock signal is therefore divided into four possible states, or four clock pulses are counted during the operational cycle of the circuit.

In binary terms, the circuit in figure 5-6 can count from 0 to 11 (or 0 to 3 in decimal). The counter starts at 00 (0), and the next clock pulse makes the count 01 (1). The next pulse makes the count 10 (2), with the following pulse incrementing it to 11 (3). That is the limit of this circuit, and the next pulse sends it back to 00.

The 74HC76 is similar to the 74HC73 except that each flip-flop has a preset input in addition to the clear input. The 74HC76 can be used in the circuits shown in figures 5-5 and 5-6 as long as both the clear and preset inputs are high. Figure 5-7 shows a circuit using two 74HC76 devices that can count from 0000 to 1111 in binary (or 0 to 15 in decimal). It is also a divide-by-16 circuit, with one output at D for every 16 clock pulses. There are intermediate count/division steps at A (divide-by-2), B (divide-by-4), and C (divide-by-8). The division/count range can be extended by adding additional flip-flops in the same manner.

Counter/dividers are available in IC form, and these contain numerous internal flip-flops in the same fashion as that shown in figures 5-5 to 5-7. There are several such devices available, and these were described in chapter 2.

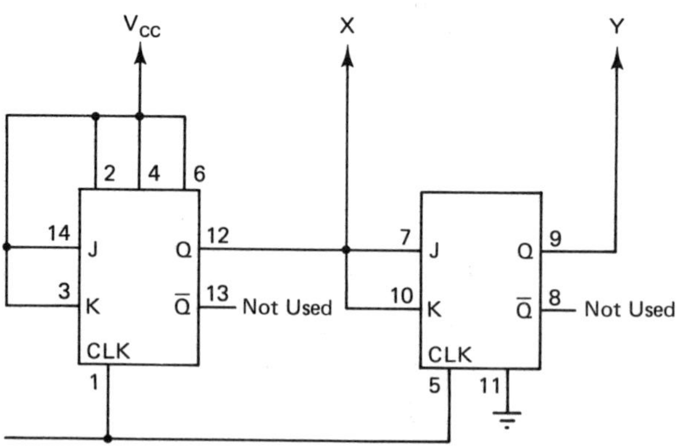

Figure 5-6 Divide-by-Four Counter Using the 74HC73

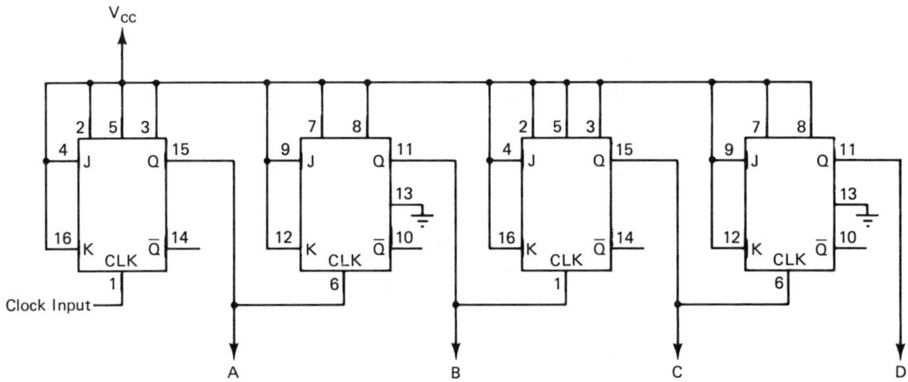

Figure 5-7 Up Counter Using the 74HC76

Several terms are used to describe counters. One is the *direction* the circuit counts in. An *up counter* starts with all outputs at 0 and goes "upward" until all outputs are high (1); a *down counter* starts with all outputs at 1 and goes "down" until all outputs are low (0). Some ICs allow counting in either direction and are called *bidirectional.* Another term is *modulo,* which means the range over which a counter counts or divides before recycling. A "modulo 10," or *decade,* counter has a range of 0 to 9 (or it can be thought of as dividing the input signal by 10). Counter/dividers can also be described as ripple or synchronized. In a *ripple counter,* the flip-flops in the counter are individually "clocked" by the output of the immediately preceding flip-flop. The clock signal can be said to "ripple" through the counter, much like ripples caused by a stone tossed into water. In a *synchronized counter,* all flip-flops are clocked simultaneously by an independent clocking signal. The major practical difference between the two types is that a ripple counter is slower than a synchronous one due to the "ripple through" time.

Counters are often described as being binary (hexadecimal) or decade (BCD). A *decade,* or *BCD, counter* is a base-10 counter that counts from 0 to 9. A *binary,* or *hexadecimal, counter* is a base-16 counter that can count from 0 to 15 (or from 0 to F in hexadecimal). A few counters with different bases (such as base-12) are also available. Some counters may be cascaded to count or divide higher than their individual range. Not all counters permit this, however.

The 74HC93 is a basic binary counter. Its internal circuitry consists of separate divide-by-2 and divide-by-8 circuits, as was shown in chapter 2. Note that each section has its own clock input. This counter can be configured to count to different levels, as shown in figure 5-8. In each case, the external clock signal is applied to the divide-by-2 counter. In turn, the output of the divide-by-2 counter serves as the clock input for the divide-by-8 input. In the divide-by-10 counter, one output of the divide-by-8 section serves as the one reset input of an AND whose output is fed to both counter sections. Another

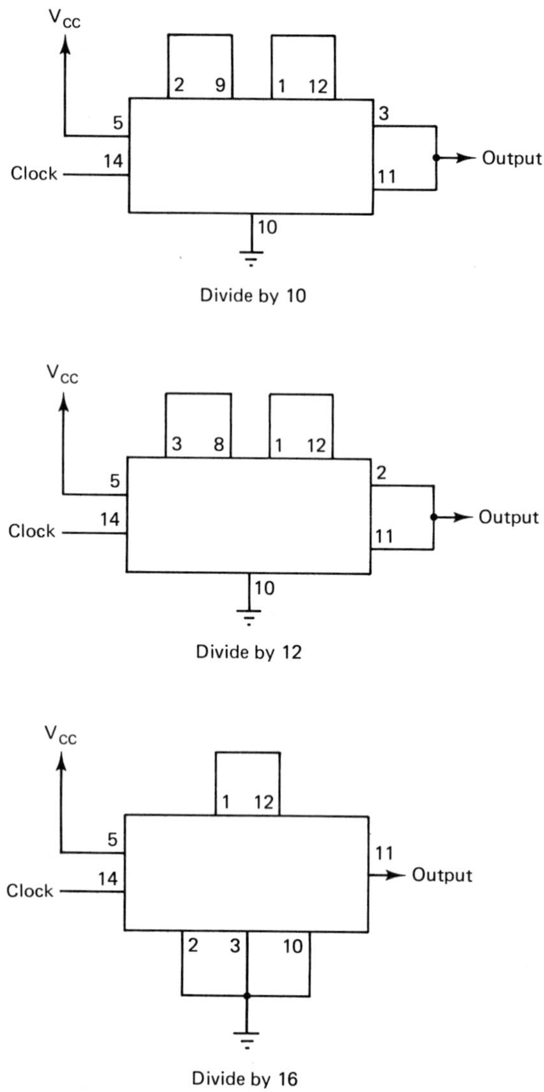

Figure 5-8 Counter Applications of the 74HC93

output of the divide-by-8 circuit (Q3) is the other reset input. When both reset inputs are high, all outputs (Q0, Q1, Q2, and Q3, located at pins 12, 9, 8, and 11, respectively) go to low. The divide-by-10 configuration therefore "aborts" the count after 9, because at that point in the counting cycle both Q1 and A3 go high and reset the entire counter. A similar approach is used in the divide-by-12 counter. In the divide-by-16 configuration, both reset inputs are tied to ground so that the counter cannot be reset. The counter can then go through a complete 0 to 15 cycle before all outputs go low again.

The 74HC93 cannot be cascaded to count beyond its maximum range. However, counters such as the 74HC4017 can. The 74HC4017 is the HCMOS equivalent of the 4017 standard CMOS decade counter and was described in detail in chapter 2. This versatile device can use positive- or negative-edge clocking and has ten separate inputs. One of the inputs goes high to represent the count while the others remain low. There is also a "carry" output that goes high when the count exceeds 10, which allows multiple devices to be cascaded. Figure 5-9 shows a 0 to 99 counter/divider constructed using two 74HC4017 devices. The switch either allows continuous counting or allows all outputs to be set to 0. The carry output of the first device (at pin 12) serves as a positive-edge clock trigger for the second device at pin 14. Once operation is under way and the count exceeds 10, there will be two high outputs: a rapidly changing one on the first device as it counts from 0 to 9, and a more slowly changing one at the second device as it advances one position each time the first device goes through a complete counting cycle.

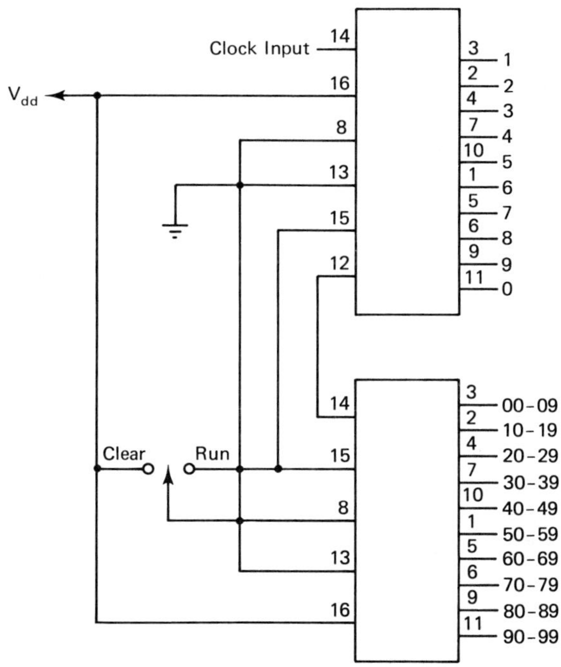

Figure 5-9 0–99 Counter Using the 74HC4017

SHIFT REGISTERS

The flip-flops making up a counter/divider can be thought of as "passing along" input pulses from one flip-flop to the other; when one flip-flop gets "full," it passes along a bit to the next flip-flop. This is the concept behind a *shift register.* This is a circuit where several flip-flops are connected in series so that the outputs of one flip-flop serve as the inputs for the next flip-flop. All flip-flops share a common clock signal, and each input pulse (or bit) into the circuit is shifted one position with each clock pulse. Shift registers can be used as memory and data storage devices or to convert data between serial (one bit after another on a single line) and parallel (each bit has its own separate line) forms.

Figure 5-10 shows a four-bit shift register that uses two 74HC76 J-K flip-flop devices. This particular type of shift register is known as a *serial input/parallel output (SIPO)* circuit, since the data enter in serial form and are taken from the circuit in parallel. This is also known as a *shift right* shift register, since the data enter the circuit at the leftmost flip-flop and gradually move toward the right. In figure 5-10, the input is directly applied to the J input of the first (leftmost) flip-flop and simultaneously through an inverter (from a 74HC04 hex inverter IC) to input K. This arrangement "splits" the single-line serial input into two complementary inputs for the flip-flop chain. The outputs making up the 4-bit parallel output (labeled A to D) are taken from the Q output of each flip-flop.

Shift registers may also be implemented using D flip-flops. Figure 5-11 shows "generic" D flip-flops being used to implement a circuit similar in function to that in figure 5-10. (For simplicity's sake, clear and preset inputs have been deleted but must be kept in mind and set to the proper logic level in "real-world" versions of this circuit.) J-K flip-flops are generally preferred for shift register applications, however, since circuits using J-K flip-flops are faster than those using D flip-flops.

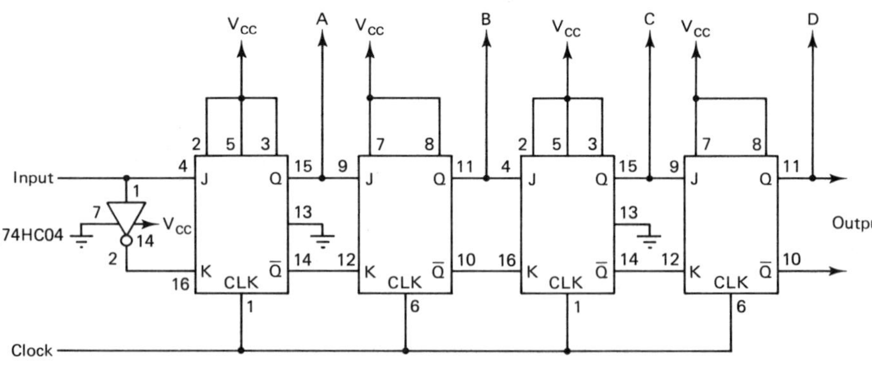

Figure 5-10 Serial Input/Parallel Output Shift Register Using the 74HC76

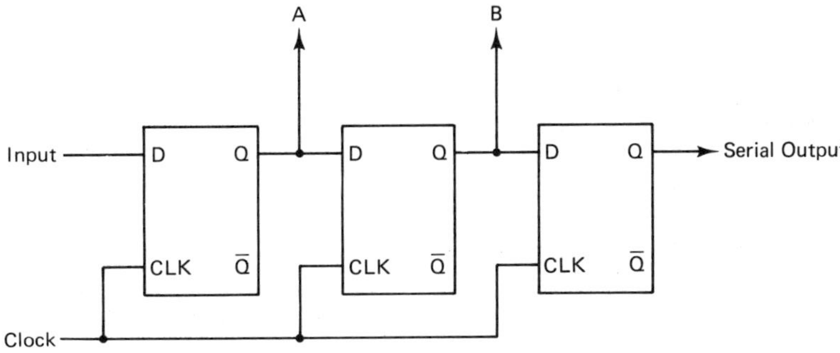

Figure 5-11 Shift Register Using D Flip-Flops

Although uncommon, it is also possible to enter data at the rightmost flip-flop of a shift register and move these data toward the left; this is known as a *shift left* circuit. Shift registers that can move data in either direction are known as *bidirectional.* In addition to SIPO shift registers, there are also *serial input/serial output* (SISO), *parallel input/serial output* (PISO), and *parallel input/parallel output* (PIPO) types. Shift register ICs that can perform these functions are available.

One example of a versatile shift register is the 74HC194 "universal" 4-bit bidirectional shift register, described in chapter 2, which can function as a SIPO or a PIPO circuit. This device can accept data in serial or 4-bit parallel form and shift the data to the right or left. A related device is the 74HC299 8-bit bidirectional shift register. This device has an 8-bit parallel data port that can be used for either input or output. There are also separate serial inputs for data that are to be shifted either left or right. The mode of operation (serial or parallel) and direction of shifting are controlled by signals applied to control input pins.

Another device described in chapter 2 is the 74HC589 8-bit SISO or PISO counter; selection of the mode of operation is via a high or low logic level control signal. There are eight parallel inputs into a data latch (latch circuits will be discussed further in the next chapter), and the data latch outputs go into a shift register. There is also a serial input directly into the shift register. The 74HC589 has a three-state output, meaning that the output can be set to a high impedance level so it is electrically isolated from any circuits that may be connected to it. This is important in data bus-oriented systems, since any connection to a bus line that draws current could disturb the proper operation of the signals distributed by the bus.

OTHER CLOCKED LOGIC APPLICATIONS

Not all clock signals need to be "real" clock signals such as those obtained from a separate clock signal oscillator. Any source of pulses can do just as well.

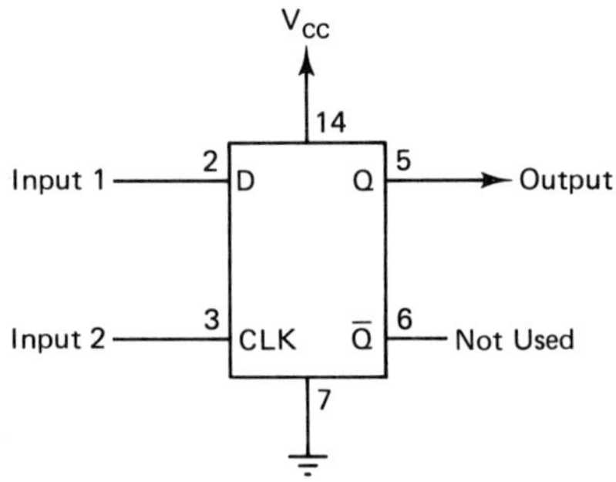

Figure 5-12 Digital Mixer Using the 74HC74

Figure 5-12 shows one section of a 74HC74 dual D flip-flop device in which two different input signals are applied. The resulting Q output is the difference between input 2 and input 1 (or input 2 minus input 1). This circuit is known as a *digital mixer.* It can be used with a LED connected to the Q output to indicate whether two digital signals are equal. If connected to light when the output is low, the LED will glow when the two inputs are unequal. If connected to light when the output is high, the LED will glow when they are equal. Since the outputs are complementary, another LED could be connected to the Q̄ output to give both actions and a "lit" indication of the frequencies of the two input signals.

Counter/dividers can be used with a variety of signal sources and combined with logic gates to produce output signal frequencies other than those normally available from HCMOS counters. Figure 5-13 shows a circuit that produces one output pulse per minute using the standard 60 Hz power line frequency as an input and reference. The 60 Hz AC signal is obtained directly from the 120 V AC line through a 1 M resistor, which is applied to one section of a 74HC14 Schmitt trigger device. The Schmitt trigger takes the AC waveform and "squares it up," producing an equivalent square wave signal having the same frequency as the AC input signal. The resulting square wave signal is applied to the 74HC4040, a binary ripple counter. Outputs Q5 (pin 5), Q10

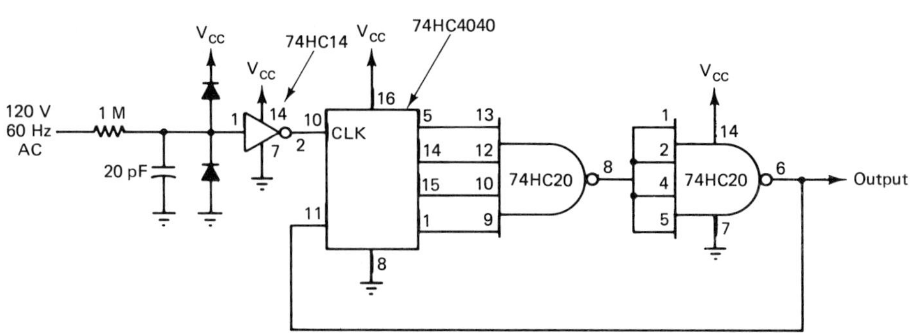

Figure 5-13 60 Hz Timebase

(pin 14), Q11 (pin 15) and Q12 (pin 1) of the counter are selected, causing a reset of the counter every 3600 input pulses from the Schmitt trigger. The counter output is applied to both sections of a 74HC20 dual 4-input NAND gate, which decodes the four counter outputs and produces a single output pulse every minute and then resets the counter by feeding back part of the output to the reset input (pin 11) of the 74HC4040. (*CAUTION!* Unless you have had some experience with building and using circuits obtaining power directly from the AC line, you should not attempt to obtain a 60 Hz signal from the AC line through a resistor.)

6

Tri-State Output Devices

The HCMOS devices discussed in chapters 3, 4, and 5 are characterized by the fact that their output(s) must assume either a high or a low logic level. There are other HCMOS devices whose outputs can be at a *high impedance* level in addition to being high or low. These are known as *tri-state output* devices.

The high impedance output state is the equivalent of an open circuit. When a tri-state device is in its high impedance state, its output level is neither high nor low logic; instead the output is effectively "disconnected" from anything to which it is connected. The high impedance state is selected by a separate input known as the *select* or *enable* output, and this output overrides any other inputs to the tri-state device. The high impedance output state makes tri-state logic handy for data bus applications. A *data bus* is a set of parallel electrical lines, each carrying a single bit of data. Real-world buses are commonly grouped into four, eight, sixteen, or other numbers of bits to reflect the size of each data word sent along them. Since tri-state devices have no effect whatsoever on a bus line when in a high impedance state, they can serve as the "glue" between the bus network and the rest of a system.

Another advantage for bus connection purposes is that several tri-state outputs can be connected together to form a single output, much like open collector HCMOS devices. Unlike open collector ICs, however, tri-state devices will have greater noise immunity and better speed.

There are several interesting tri-state devices available in HCMOS. In this chapter, we will examine some of them and their applications; other tri-state output devices covered in chapter 2 will follow the same principles as those outlined here.

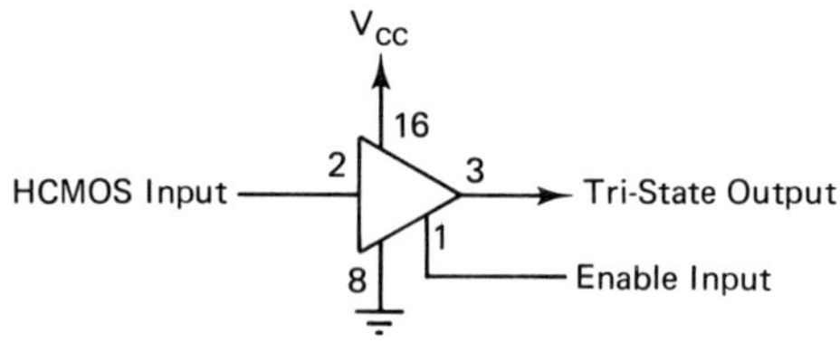

Figure 6-1 Using the 74HC367 to Add Tri-State Outputs

BASIC TRI-STATE DEVICES

The simplest type of tri-state output device is a buffer or inverter with a select or enable input in addition to the normal single input. Two such devices are the 74HC367 hex tri-state noninverting buffer and the 74HC368 hex inverting buffer. Details of these were given in chapter 2, but it is useful here to note that *both have separate two-bit and four-bit sections.* Both devices have two independent enable inputs, one controlling two buffers or inverters and the other controlling the remaining four buffers or inverters. The 74HC367 and 74HC368 function as normal buffers or inverters while the appropriate enable signal is low. When it is high, the outputs are at a high impedance state. The two separate enable inputs of these devices means that they may have some outputs at high impedance while the others can operate in a normal manner.

Figure 6-1 shows how these devices can be used to add a tri-state output to those HCMOS devices that lack it. In the example shown, one buffer of a 74HC367 is used to provide a noninverting tri-state output. If an inverted output is desired, the 74HC368 can be used.

Gates with tri-state outputs can be used in selection circuits where the ultimate output only assumes a "normal" high or low output. The circuit in figure 6-2 uses two buffers from a 74HC367 and one inverter from a 74HC04

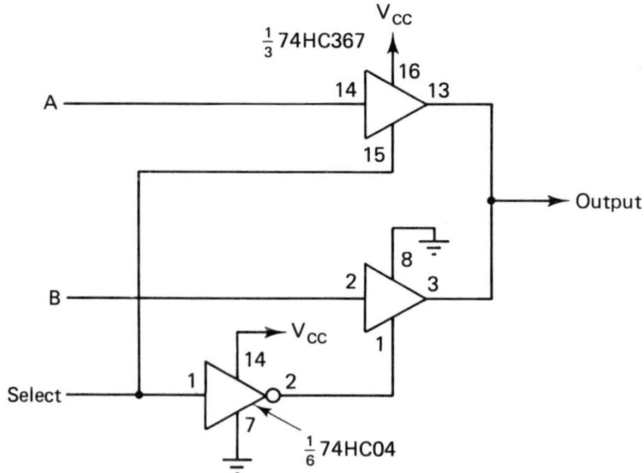

Figure 6-2 1 of 2 Data Selector Using the 74HC367

to select one of two input signals, labeled A and B, and directs one of them to the output depending on the signal present at the select input. If the select input is low, then the signal at A appears at the output; if it is high, the signal at B appears at the output.

As mentioned before, a common application of tri-state devices is in circuits connected to a common data bus. Figure 6-3 shows two 74HC367 devices used to read from or write onto a four-bit bidirectional data bus. This bus is called *bidirectional* because data may flow "up" or "down" the bus, as indicated by the arrows. The "write" and "read" inputs for the two devices are their four-bit enable inputs. When both enable signals are low, data can flow from the device on the left onto the bus, where the data are "received" and output to other circuitry by the 74HC367 on the right. If either enable signal (or both) is high, that 74HC367 is "isolated" from the bus and data cannot be

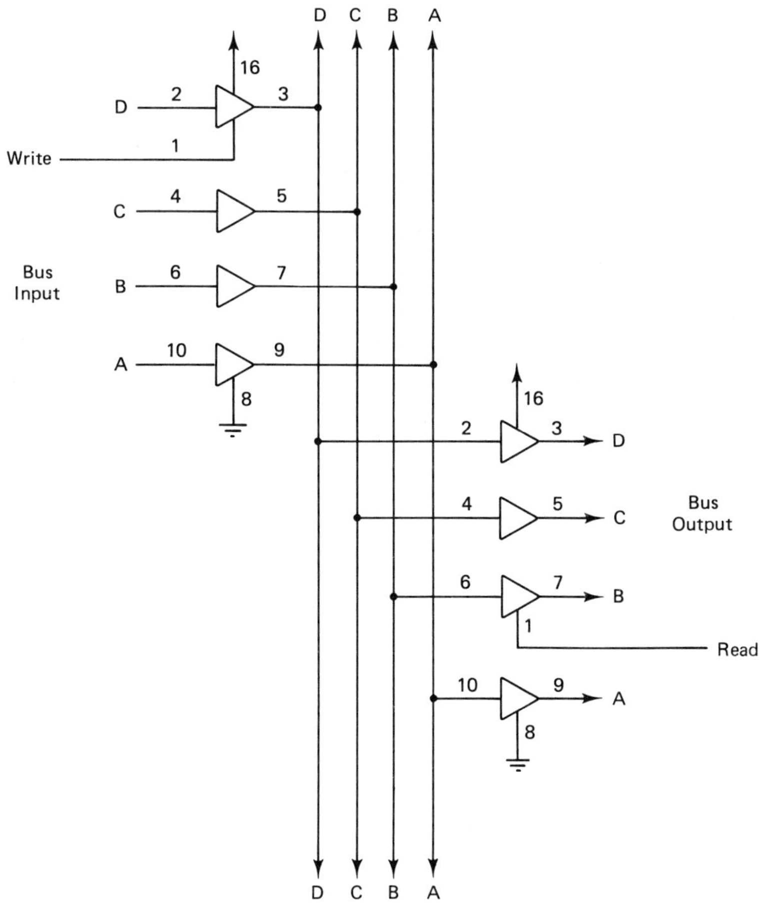

Figure 6-3 Using Two 74CH367 Devices in a Bidirectional Data Bus

input to or read from the bus. The two devices are independent of each other, and the status of one can now affect the status of the other (unless, of course, they both share a common enable signal). There is also no need for one bus "transmitter" for each bus "receiver." There could be a dozen transmitter 74HC367 devices connected to the bus, but only one receiver 74HC367. However, only one device may transmit at any given moment; there is no limit on the number of devices simultaneously receiving.

BUS INTERFACING DEVICES

An arrangement such as that shown in figure 6-3 is not very practical for most real-world applications. However, the basic principles carry over to other devices suitable for bus interfacing and communication, including some specifically designed for such purposes.

Figure 6-4 is a "restatement" of the circuit in figure 6-3 using the 74HC244 octal tri-state noninverting buffer device. This device contains two four-bit sections, with each four-bit section having its own independent enable input. In figure 6-4, the read or write function to the bidirectional data bus is selected by holding the desired select input low while the other is kept high. (Both enable inputs may be low, but then the read and write functions would be taking place simultaneously and there would be no need to place the data on the bus; a direct connection between the circuits would be best.) A similar circuit may be implemented using the 74HC240, which is an inverting buffer device.

In some cases, two-way communication along a bus is not necessary. We may want to simply allow signals to move from one bus to another in a single direction at certain times but not others. This circuit is sometimes known as a *bus buffer*, and an example using the 74HC244 is shown in figure 6-5. This

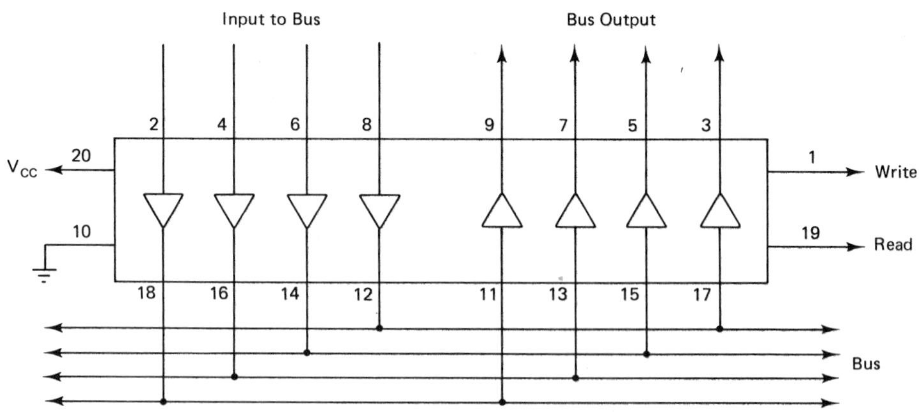

Figure 6-4 Four-Bit Bus Interface Using the 74HC244

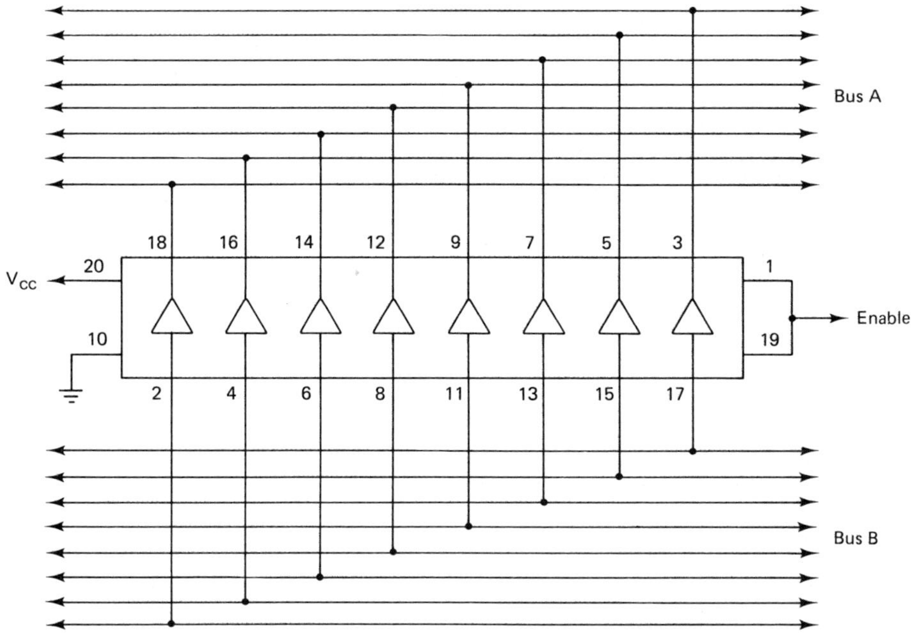

Figure 6-5 Bus Buffer Using the 74HC244

circuit serves as a "drawbridge" between two eight-bit bidirectional buses labeled A and B. The two enable inputs are tied together to form a single input, while all buffer inputs are connected to bus B and all buffer outputs are connected to bus A. When the enable input is high, the two buses are isolated from each other. If the enable input is low, data can move from bus B to bus A. An inverting version of this circuit can be implemented using the 74HC240.

BUS TRANSCEIVERS

Buffer devices such as the 74HC244 and 74HC367 are seldom the optimum solution to bus interfacing problems. Most bus systems today are at least eight bits; moreover, we usually need to be able to transfer data in either direction between the two buses. To meet these needs, *bus transceiver* devices have been developed and are available in HCMOS. A typical example is the 74HC245, which was fully described in chapter 2. This device has a *direction input* at pin 1 in addition to an enable input. As its name implies, the direction input controls which bus does the "transmitting" and which is the "receiver." When the direction input is high, data flow from bus A to bus B; when it is low, data flow from bus B to bus A. The direction input only has an effect when the enable input (pin 19) is low. If it is high, all inputs/outputs have a

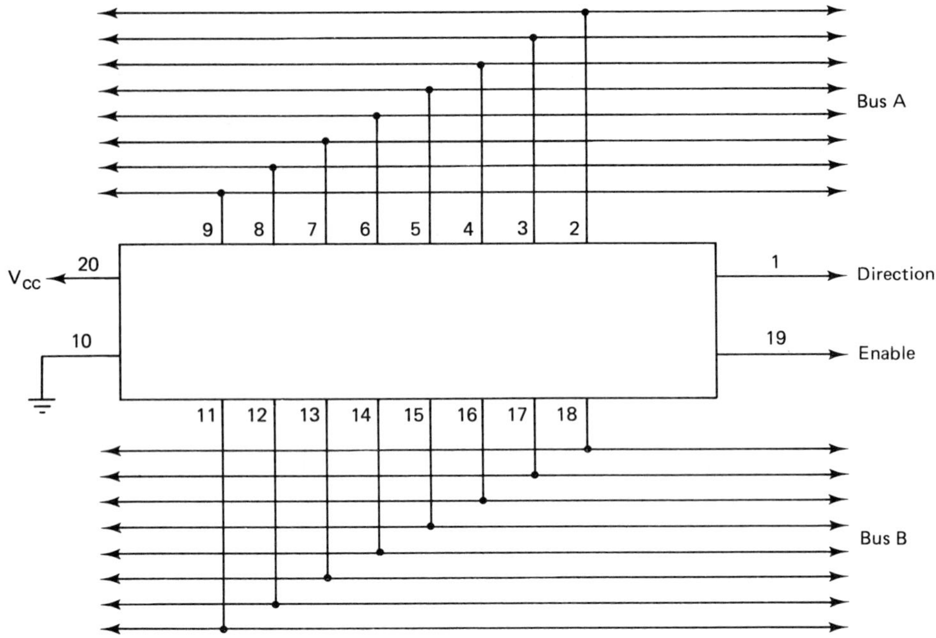

Figure 6-6 74HC245 Bus Transceiver Application

high impedance level and the buses are isolated from each other. Figure 6-6 shows an application of the 74HC245 with two eight-bit bidirectional buses.

BUS STORAGE REGISTERS

In chapter 5, we discussed types of HCMOS devices known as shift registers, which can store and manipulate data. A *register* is a device or circuit that stores data. In bus systems, it is frequently necessary to have storage devices for data placed on or read from the bus network. A bus register can be implemented using either flip-flop or latch devices.

One example of a latch-based bus register uses the 74HC373, as shown in figure 6-7. The basic operation and logic table of this device were discussed in chapter 2. The key to this device's operation is in the latch enable input at pin 11. When the latch enable input is high (and the output enable at pin 1 is low), the outputs follow any changes in the inputs. When the latch enable input goes low, however, the outputs become latched at their current logic state and all changes in the input are ignored, and the device can be said to "store" a given input. If the output enable goes high, the outputs go to a high impedance. However, the data stored by the 74HC373 in its high impedance state are not altered by making the output enable input high. If it is returned to low, the stored data will appear once more at the outputs.

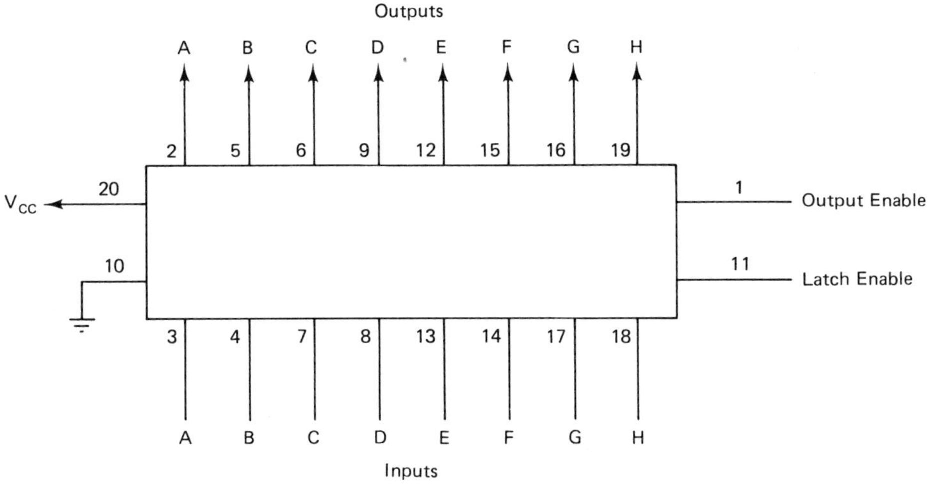

Figure 6-7 Using the 74HC373 as a Storage Register

A variation of figure 6-7 is shown in figure 6-8. This circuit uses the 74HC374 D flip-flop IC to implement a *clocked* bus register. For normal operation, the output enable (pin 1) must be kept low. During normal operation, data at the inputs are loaded into the register on the rising edge of each clock input pulse at pin 11. These data then appear at the outputs of the 74HC374. However, when the clock input is at a steady high or low level, the outputs do not change regardless of any changes in the input. As with the 74HC373 circuit in figure 6-7, a high output enable signal will set the outputs to a high impedance level but will not affect the contents of the register. *In addition, data can be entered into the register with each clock pulse even though the out-*

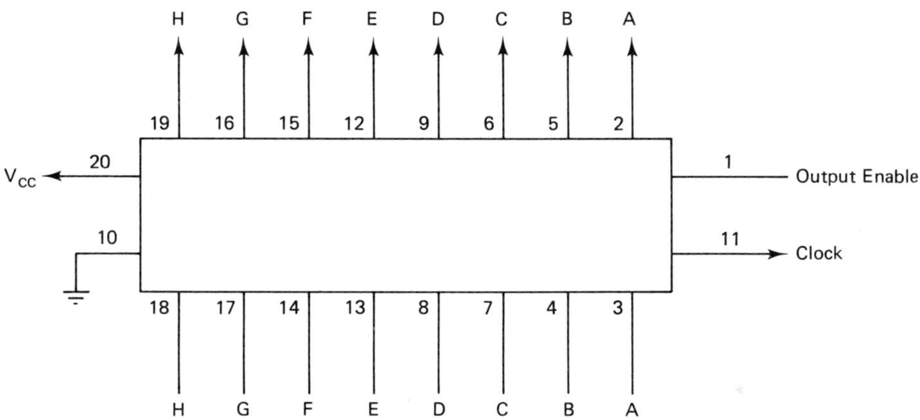

Figure 6-8 Using the 74HC374 as a Closed Register

puts are at high impedance. While a high impedance output state does not alter the contents of the register, it does not preserve them either.

Figure 6-9 shows how two bus registers (two 74HC373 devices in this case) can be used on an eight-bit bus network. Note that pin 1 of each device has been relabeled "connect/disconnect," since this more accurately reflects the function of pin 1 in this configuration. With register 1, a low connect/disconnect input will connect register 1 to the bus. When the enable 1 input is high, the signals at the inputs appear at the output of register 1 and are placed on the data bus. If enable 1 is low, the register is latched and all further changes in the inputs are ignored; a constant signal is presented to the bus. The usual practice when the register is latched is to isolate it from the bus by making the signal at pin 1 high. In the case of register 2, the connect/disconnect 2 must be low for it to read signals on the bus and make them appear at the outputs. If the signal is low, register 2 is isolated from all the following circuitry. The enable 2 input should be kept high to enable the outputs of register 2 to follow changes in signals on the bus. If enable 2 goes low, register 2 latches and stores its last set of inputs from the bus. Only one register can write to the bus at any time, but any number of registers can read from the bus.

Figure 6-10 shows a single 74HC374 device used as a common input/output register. It has been configured to provide "throughput" from its D inputs to its Q outputs. Pin 1 has been relabeled "read/write" to reflect its function accurately in this configuration. When the signal at pin 1 is high, data on the bus are loaded into the 74HC374 on the rising edge of a clock pulse. (Remember, data can still be loaded even though the device is in its high impedance state.) When the pin 1 signal is low, the data in the device are written onto the bus network.

Configurations similar to figures 6-9 and 6-10 can be implemented using other tri-state latches and flip-flops.

OTHER TRI-STATE LOGIC DEVICES

Several data selector/multiplexer devices are available with tri-state outputs. These have only limited applications in bus networks and are primarily used where we wish to combine several device outputs into a single output. If a tri-state output is needed from other types of HCMOS devices, then a circuit similar to that shown in figure 6-1 should be used. Given the need for an additional control signal to select high impedance or normal operation, and the possibility of accidentally setting an output to high impedance and causing abnormal circuit operation, tri-state logic outputs should only be used in those situations where ordinary HCMOS devices will not do the job, such as connections to bus networks.

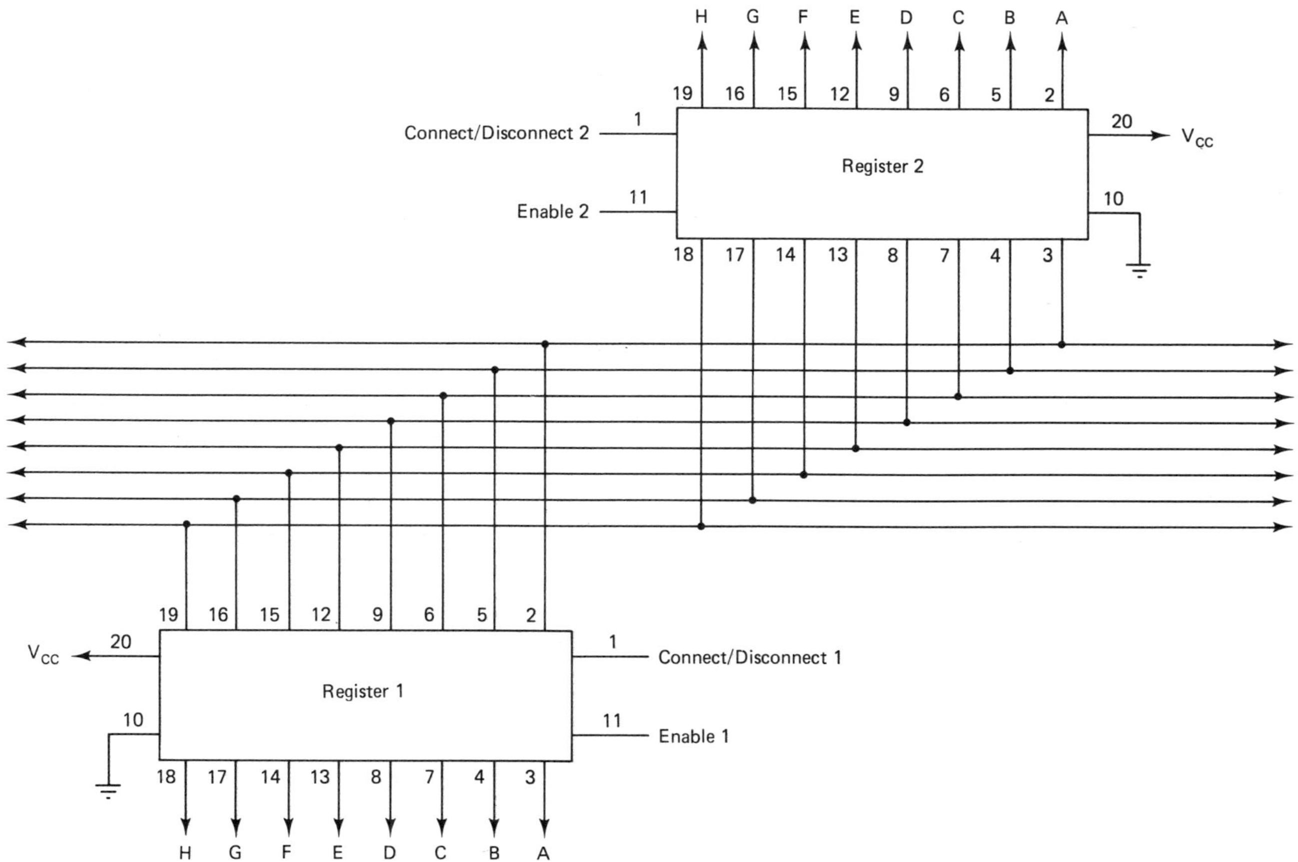

Figure 6-9 Using the 74HC373 Devices for Bus Communications

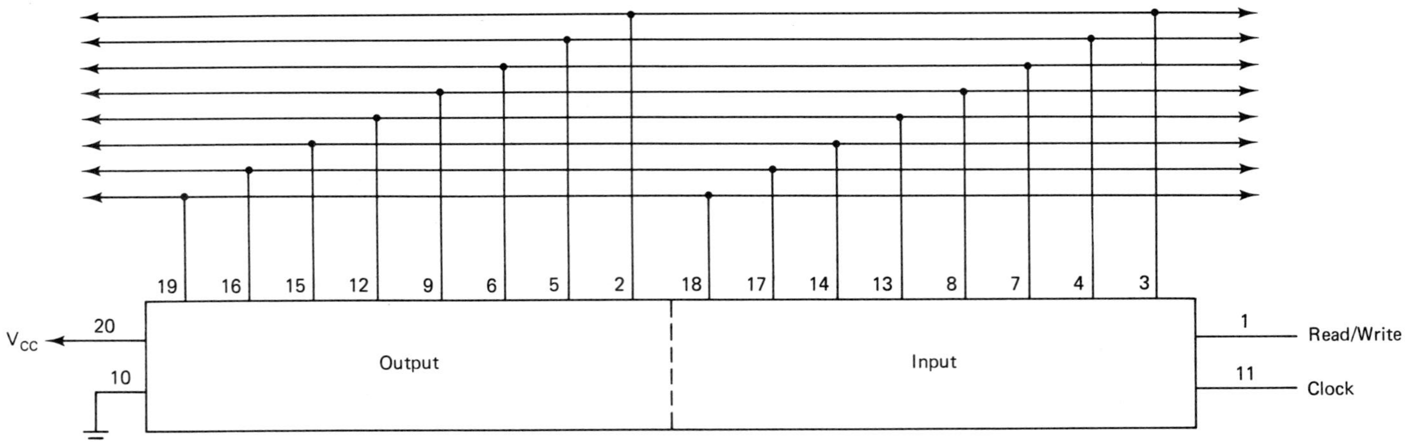

Figure 6-10 Using the 74HC374 in a Common Input/Output Register Configuration

7

Miscellaneous HCMOS Devices

The rich variety of available HCMOS devices means that several do not fit into any of the device types covered in previous chapters. In this chapter, we will look at such miscellaneous devices, giving special attention to HCMOS equivalents of standard CMOS ICs.

THE 74HC4538 DUAL MONOSTABLE MULTIVIBRATOR

A *monostable* (or *one-shot*) *multivibrator* is a circuit that emits a single output pulse of a fixed duration when an external *trigger* signal is received. (This is in contrast to the astable multivibrator, which emits a stream of pulses.) Most monostable multivibrators are built around a timer IC, such as the 555 or its functional equivalents. However, the 74HC4538 packs two independent monostable multivibrators with dual trigger inputs into a single package. It is the HCMOS equivalent of the 4538 standard CMOS device, and it requires only an external timing capacitor and resistor for operation. The output pulse duration can range from less than 1 μs to well over a full second.

This device was described in detail in chapter 2. Figure 7-1 shows a typical application for one "half" of the 74HC4538. The circuit has complementary outputs, Q and $\bar{Q}$, and two trigger inputs labeled A and B. There is also a reset input at pin 3. The reset input must be held to a high logic level for normal operation; making it low will cause Q to be low and $\bar{Q}$ to be high. This monostable multivibrator can be triggered in two ways: by applying a positive-

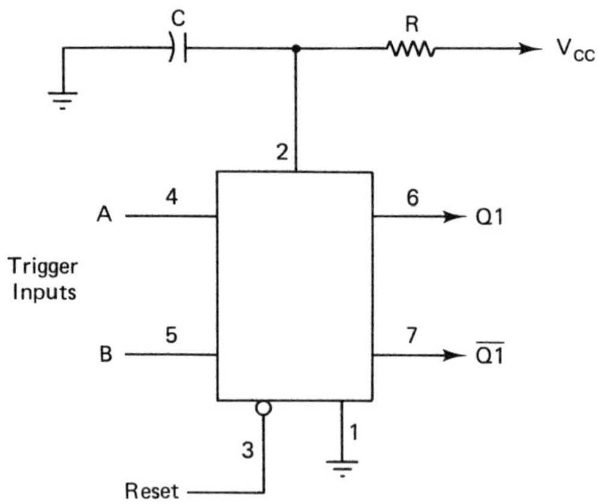

Figure 7-1 74HC4538 Monostable Multivibrator Application

edge signal to input A while input B is high, or by holding input A low while applying a negative-edge pulse to input B. All other combinations of inputs will not trigger the circuit.

Once triggered, the output of Q will be a positive pulse while $\bar{Q}$ will have its complement, or a negative pulse. The duration of the output pulse can be found by the formula

$$\text{output pulse duration} = 0.5 * R * C$$

where the maximum value of C should not exceed 10 μF and the maximum value of R should not exceed 1 M. Once triggered, the circuit must be retriggered to produce an output pulse. This is done by making both trigger inputs receive a low signal, after which another positive- or negative-edge signal will produce another output pulse.

Although a useful device for some applications, the 74HC4538 can be a difficult circuit to work with, especially at the extremes of its output pulse duration range. Precision value components are necessary for accurate output pulse times, as is careful placement of the timing components. Because of this "touchiness," many prefer to use conventional 555 or equivalent timer devices.

THE 74HC280 PARITY GENERATOR/CHECKER

Parity checking is a method used with computer or microprocessor systems to detect errors in the recording or transmission of data. This method consists of adding up the number of bits in a unit of data (usually a byte) and comparing it with another bit, called the *parity bit,* which is sent as part of the data unit. The number of bits in the data unit will either be odd (1, 3, 5, 7) or even (0, 2, 4,

6, 8), and the parity bit itself will be 0 (even) or 1 (odd). The bit count is compared with the parity bit, and if they match, the data unit is considered to have been successfully recorded or sent. If they differ, an error is assumed to have occurred. The parity bit usually precedes any data unit and is sent along with it; in "100100101" the first one is the parity bit. In this case, the number of bits is positive (four) while the parity bit is odd, so the data unit is assumed to have suffered an error in transmission or reception.

Parity checking is greatly simplified through the use of the 74HC280 parity checker and generator. Figure 7-2 shows a typical application for this device. The 74HC280 is known as a parity generator because it has two outputs that go high or low depending on the parity of the nine-bit signal applied to inputs A through I. Input A (pin 8) is considered the parity bit, with inputs B through I constituting the byte. If the number of inputs that are high is an even number (0, 2, 4, 6, 8), the even parity output goes high while the odd parity output goes low. If the number of inputs that are high is odd (1, 3, 5, 7, 9), then the odd parity output is high while the even parity output is low.

It is often necessary to perform parity checking on data units consisting of two or more bytes plus a parity bit. Fortunately, the 74HC280 can easily be cascaded, as shown in figure 7-3, to check larger data units. As shown, the odd parity output of the first device is connected to the parity bit input of the second device while the even parity input of the first device is not used. Additional devices can be cascaded as shown here to check larger data units (such as 32 bits).

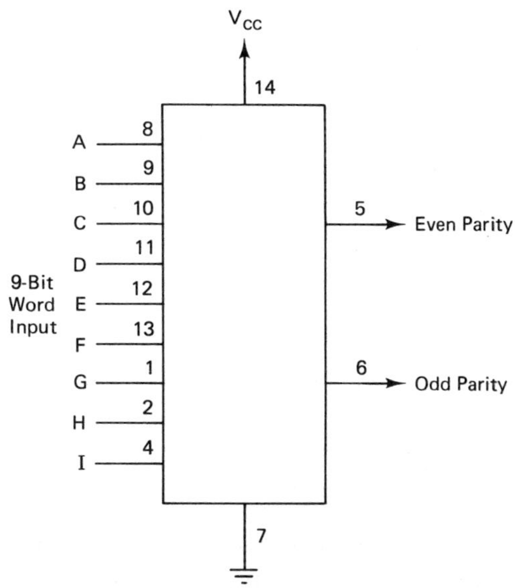

Figure 7-2 74HC280 Parity Generator/Checker Application

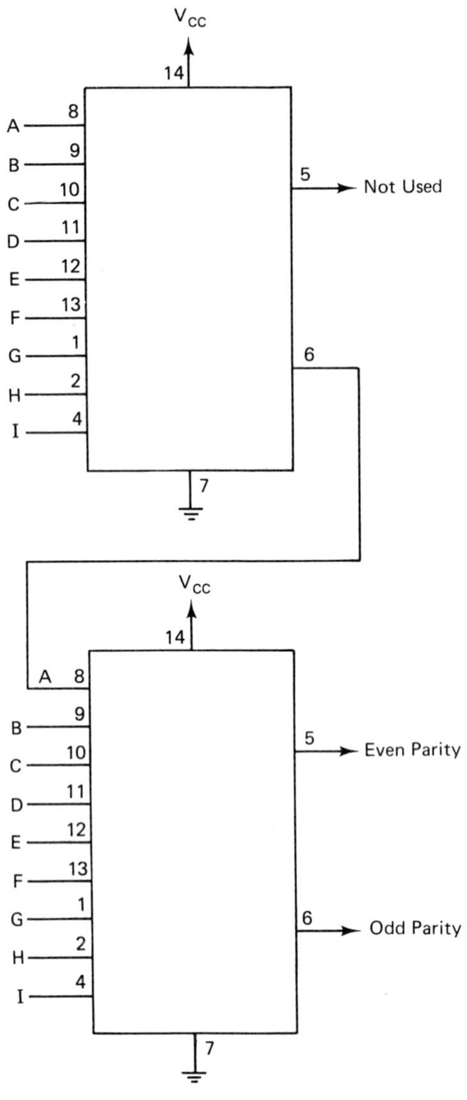

Figure 7-3 Cascaded 17-Bit Checker Using the 74HC280

EQUALITY AND MAGNITUDE COMPARISON

In many situations, it is necessary to do some type of comparison between two data units of the same size. A common need is to know whether two data units are equal. The 74HC688 eight-bit equality comparator device greatly simplifies this task. Figure 7-4 shows how it takes two bytes, labeled A and B, at its inputs and indicates their relationship at the A = B output. If both bytes are equal, the A = B output goes low; otherwise it remains high. There is also a

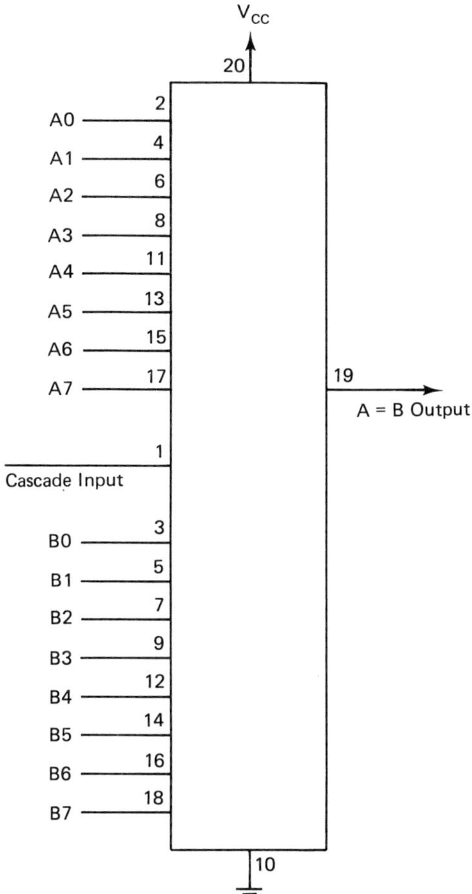

Figure 7-4 74HC688 8-Bit Equality Comparator Application

cascade input at pin 1 which should be kept low for normal operation of the circuit in figure 7-4. If the cascade output is high, the output will remain high regardless of the actual equality or inequality of the input bytes. Data units larger than a single byte can be compared by cascading two or more 74HC688 ICs. This is done by connecting the A = B output of the first device to the cascade input of the second, and so forth, as additional devices are cascaded. The first 74HC688 will compare the most significant bits in the data unit, while the last device in the cascade chain will compare the least significant bits. The A = B output of the last IC will reflect the equality or inequality of the entire two data units.

It is more likely, however, that we would want to know whether two data units are equal and, if they are not, which one is greater. This type of *magnitude comparison* is possible with the 74HC85 four-bit magnitude comparator device. Figure 7-5 shows how this IC is used to compare two nibble inputs. The 74HC85 has three cascading inputs labeled A > B, A = B, and A < B.

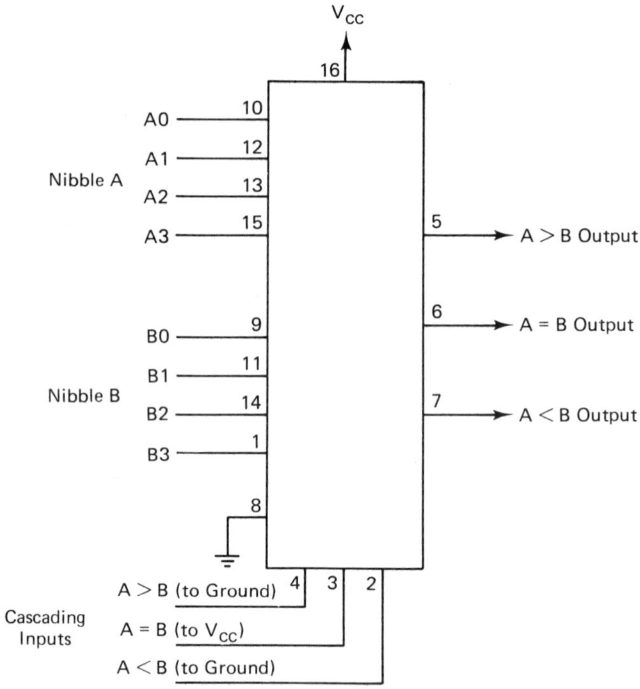

Figure 7-5 74HC85 4-Bit Magnitude Comparator Application

For normal operation of the circuit in figure 7-5, the A > B and A < B inputs must be kept low while the A = B input must be kept high. When this is done, the IC will compare input nibbles A and B and will make the appropriate output high while the other two will remain low.

Two 74HC85 devices may be cascaded to compare two data units of a byte in length. Figure 7-6 shows how this can be done. Note that the outputs of the first device are connected to the corresponding cascading inputs of the second. The cascading inputs of the first device are "tied" to ground or V_{cc}, as indicated in figure 7-5. The first 74HC85 compares the least significant bits of each data unit, while the second compares the most significant bits. Additional devices may be cascaded in this manner to compare even larger data units.

ANALOG MULTIPLEXERS/DEMULTIPLEXERS

The basic concepts of multiplexers and demultiplexers were discussed in chapter 3. Those circuits could only handle digital signals. However, HCMOS provides several multiplexer/demultiplexer devices that can handle either analog or digital signals and voltages. These devices are also "bidirectional" in

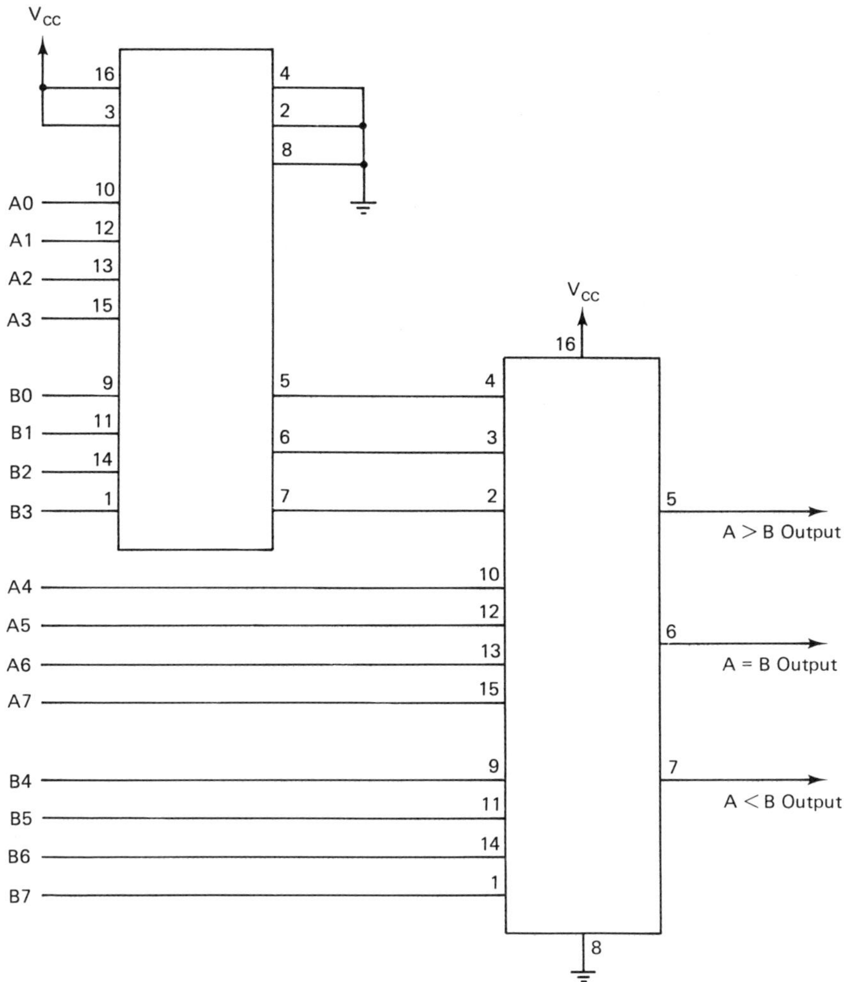

Figure 7-6 Comparing 8-Bit Words with Cascaded 74HC85 Devices

the sense that they have pins that can serve as either inputs or outputs; these are known as *input/outputs.* One popular such IC is the 74HC4051, which is a single-section, eight-position, and common off design. This chip can operate in an analog mode using positive (V_{cc}) and/or negative (V_{ee}) supply voltages over a range from 2 to 12 V. It can also be used with digital signals by operating it with an HCMOS power supply (V_{cc} of +5 V and reference to ground rather than a negative polarity voltage). The 74HC4051 has eight independent input/output pins (pins 1, 2, 4, 5, 12, 13, 14, and 15) and one common input/output (pin 3). There are three address inputs that are used to select a desired input/output and an enable input (pin 6), which must be kept low for normal operation.

Figure 7-7 shows the 74HC4051 configured as a demultiplexer (or data selector). One of the eight inputs (labeled 0 through 7) will be selected and will appear at the single output according to the inputs at pins 9, 10, and 11. (The function table for this device was given in chapter 2 and will not be repeated here.) If a dual polarity power supply is not used, pin 7 (V_{ee}) should be connected to ground of the input circuit or to the ground connection at pins 6 and 8 if ground is common.

Figure 7-8 shows the 74HC4051 used as a multiplexer (or data distributor). Much as in figure 7-7, the three address inputs determine to which of the eight outputs the single data input signal is sent. The V_{ee} pin must be connected to the appropriate ground (as mentioned previously) if a single polarity power supply is used.

In addition to the 74HC4051, two other HCMOS analog multiplexers/demultiplexers are the 74HC4052 and 74HC4053. The 74HC4052 consists of two four-position sections with common off, while the 74HC4053 consists of three two-position sections with common off. Full details on these two devices can be found in chapter 2.

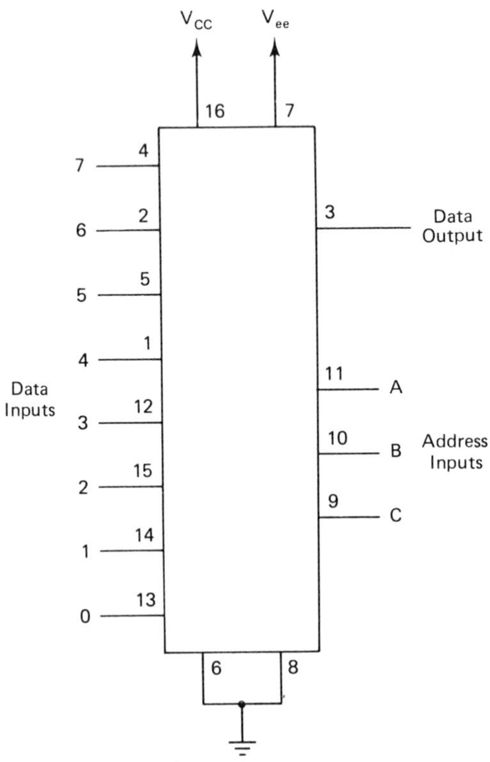

Figure 7-7 Demultiplexer Using the 74HC4051

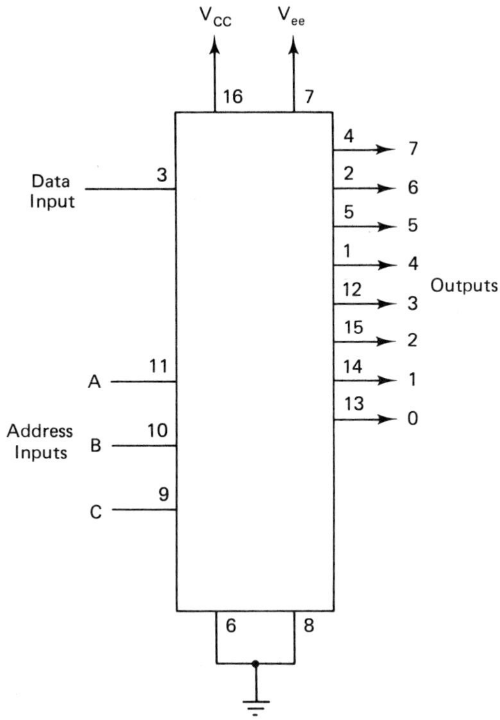

Figure 7-8 Multiplexer Using the 74HC4051

ANALOG SWITCHES

An HCMOS equivalent of a popular standard CMOS IC is the 74HC4066 quad analog switch device. As with the 74HC4051, this device can be used with either analog or digital signals and has bidirectional input/output pins. Each of the four switches has its own independent on/off control input. When the corresponding on/off input is high, the switch is on; when it is low, the switch is off. A complete description of this device was given in chapter 2. Figure 7-9 shows how it is used as a demultiplexer. There are four inputs, while the four corresponding outputs have been "tied" together to form a single output. The data select inputs are actually the on/off control inputs for the four switches. The desired input is made high, while the others are made low. If more than one control input is high, the resulting output will be a combination of more than one input signal. This circuit can easily be altered (or more descriptively, "reversed") to form a multiplexer.

Another application of the 74HC4066 is as a bus transmitter or receiver. Figure 7-10 shows how this can be done for a four-bit data bus system. This circuit can be used to place data onto a bus (outputs connected to the bus network) or to receive data from a bus (inputs connected to the bus network). Note that all the on/off control inputs are connected together to form a single

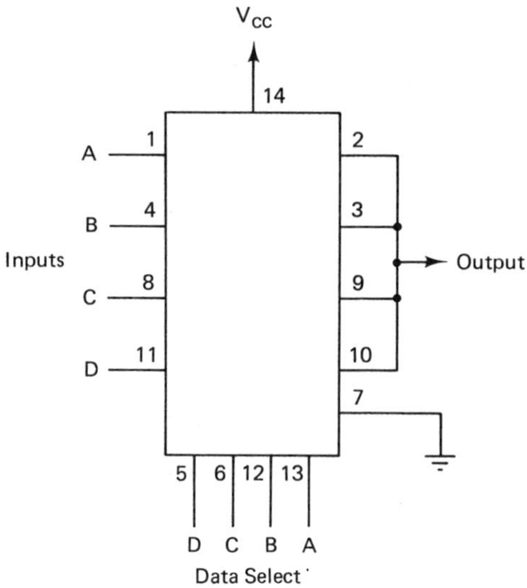

Figure 7-9 Data Selector Using the 74HC4066

control input. This circuit can be altered for non-bus system applications by separating the control inputs and using the device to send high or low signals to other digital IC inputs. Of course, perhaps the most useful application of the 74HC4066 is to allow digital control and switching of analog signals. For example, the gain of an operational amplifier circuit may be controlled by switching in different values of feedback resistors using the 74HC4066. This type of circuit is sometimes known as a "programmable gain" amplifier. In a similar fashion, the output frequency of an astable multivibrator built around a 555 timer IC or similar device could be controlled by switching different values of timing resistors or capacitors.

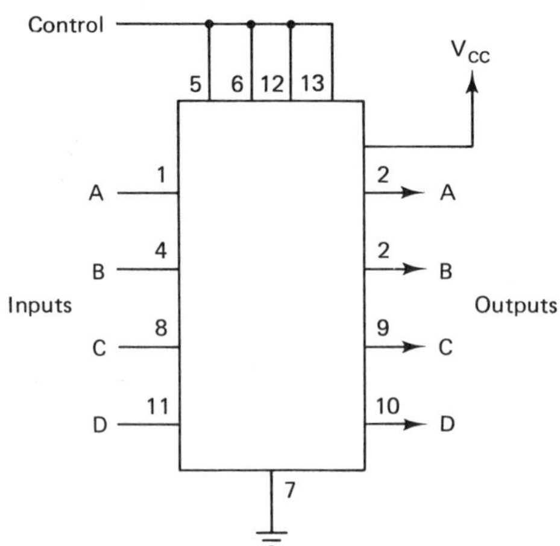

Figure 7-10 Bus Control Configuration Using the 74HC4066

Index

A

Analog multiplexers/demultiplexers, 204–6
Analog switches, 207–8
Amplifiers:
  audio/tone, 170–71
  linear, 170
  unity gain, 170–71
Asynchronous circuits, 152

B

Bus buffers, 192–93
Bus transceivers, 193

C

Clocking methods:
  edge, 173–74
  level, 173–74
  transition between states, 174
  using timer ICs, 175

Combinational logic, 151
Complimentary metal oxide semiconductor (CMOS) devices:
compared to HCMOS, 6–7
equivalent schematic of, 3
interfacing to HCMOS, 10–11
operating parameters, 4–5
susceptibility to static discharge, 5
Counter/dividers:
automatically recycling, 179
bidirectional, 181
binary, 181
cascading, 180
decade, 181
divide by two, 179
divide by four, 180
divide by ten, 182
divide by twelve, 182
divide by sixteen, 182
down counter, 181
hexadecimal, 181
ripple, 181
up counter, 181
0 to 99, 183

D

Data selectors, 190–91
Demultiplexer, 164–65
Digital mixer, 186

E

Enabled AND gate, 159–60
Equality comparator, 202–4

F

Fanout, 3
Flip-flop circuits:
D type, 158–59

definition of, 155
enabled R-S type, 157–58
J-K type, 176–77
R-S type, 157
T type, 178

H

Half adder, 162
High-speed (HC) CMOS devices:
advantages of, 1
compared to CMOS, 6–7
handling of, 7–8
interfacing to CMOS, 10–11
interfacing to TTL, 9–10
LED outputs for, 11
operating requirements, 8
packaging, 12–13
power supplies for, 8–9
testing devices for, 12
types of, 5–6

I

Input exander, 161
Input protection circuits for HCMOS, 7
Interfacing:
HCMOS to TTL, 171–72
HC to external devices, 171–72
high voltage CMOS to low voltage HCMOS, 10
TTL to HC, 10
TTL to HCT, 9–10
TTL to low voltage HCMOS, 11

L

Logic gates:
AND gates, 153
NOR gate, 153
OR gate, 153
symbols for, 152

Logic gates (*cont.*)
XNOR gate, 154
XOR gate, 153

M

Metal oxide semiconductor transistors (MOSFETs), 3–4
Multiplexer, 163–64
Multivibrator:
astable 167–68
dual monostable, 199–200
monostable, 160

O

Open drain devices, 12
Operating requirements for HCMOS, 8
Oscillators:
crystal controlled, 168–69
pulse and ramp generation, 169–70
using unbuffered inverters, 167–68
voltage controlled, 169

P

Parity generator/checker, 200–201
Power supplies, 8–9
Propagation delay, 3
Pulse stretcher, 160

S

Sequential logic, 151
Shift registers:
parallel input/parallel output, 185
parallel input/serial output, 185
serial input/parallel output, 184
serial input/serial output, 185
universal, 185

Synchronous circuits, 152

T

Timebase, 60 Hz, 186–87
Tri-state output devices:
  basic applications of, 190–92
  bus interfacing devices, 192–93
  bus storage registers, 194–98
  bus transceivers, 193–94
  definition of, 189
Transistor-transistor logic (TTL) devices:
  current consumption of, 3
  equivalent schematic of, 2
  interfacing to HCMOS, 9–10
  operating parameters, 2

U

Unbuffered devices, 167–72